Dump Philosophy

Also Available from Bloomsbury

Dust, Michael Marder
Searching for the Anthropocene: A Journey into the Environmental Humanities, Christopher Schaberg
The Green Light: A Self-Critique of the Ecological Movement, Bernard Charbonneau
Philosophical Posthumanism, Francesca Ferrando
Posthuman Glossary, Rosi Braidotti and Maria Hlavajova

Dump Philosophy

A Phenomenology of Devastation

Michael Marder
with artworks by Anaïs Tondeur

BLOOMSBURY ACADEMIC
LONDON • NEW YORK • OXFORD • NEW DELHI • SYDNEY

BLOOMSBURY ACADEMIC
Bloomsbury Publishing Plc
50 Bedford Square, London, WC1B 3DP, UK
1385 Broadway, New York, NY 10018, USA

BLOOMSBURY, BLOOMSBURY ACADEMIC and the Diana logo are trademarks of Bloomsbury Publishing Plc

First published in Great Britain 2021

Author © Michael Marder, 2021

Photography © Anaïs Tondeur, 2021

Michael Marder has asserted his right under the Copyright, Designs and Patents Act, 1988,
to be identified as Author of this work.

For legal purposes the Acknowledgments on p. ix constitute an extension
of this copyright page.

Cover design by Charlotte Daniels
Cover image: *Drowning in*, Anaïs Tondeur, 2018-20, Pigment print on Murakumo paper,
42x63cm

A catalogue record for this book is available from the British Library.

Library of Congress Cataloging-in-Publication Data

Names: Marder, Michael, 1980- author. | Tondeur, Anais, artist.
Title: Dump philosophy : a phenomenology of devastation /
Michael Marder with artworks by Anais Tondeur.
Description: New York : Bloomsbury Academic, 2020. | Includes
bibliographical references and index.
Identifiers: LCCN 2020032497 (print) | LCCN 2020032498 (ebook) |
ISBN 9781350170599 (hb) | ISBN 9781350170605 (pb) |
ISBN 9781350170612 (epdf) | ISBN 9781350170629 (ebook)
Subjects: LCSH: Environmental degradation. | Nature–Effect of human beings
on. | Environmentalism–Philosophy.
Classification: LCC GE140 .M38 2020 (print) | LCC GE140 (ebook) | DDC 304.2/8–dc23
LC record available at https://lccn.loc.gov/2020032497
LC ebook record available at https://lccn.loc.gov/2020032498

ISBN: HB: 978-1-3501-7059-9
PB: 978-1-3501-7060-5
ePDF: 978-1-3501-7061-2
eBook: 978-1-3501-7062-9

Typeset by Deanta Global Publishing Services, Chennai, India

To find out more about our authors and books visit www.bloomsbury.com
and sign up for our newsletters.

for David Buckel, in memoriam

Contents

acknowledgments

This book has been in the making over the last five years, even as the world has been disintegrating at an accelerating pace. Some of the textual materials gathered here have been published, in an embryonic form, elsewhere. An earlier version of "The Portrait of a Thing as Its Own Wastebasket" appeared as "Being Double" in *Cabinet Magazine*'s special issue on containers (Nr. 60, Winter 2015–16). "Our Polluted Senses" was initially published under the same title in *The New York Times*' "The Stone" section on October 9, 2017. A version of Chapter 16 was published as "The Writing Dump" in *American Book Review*, 39(5), July/August 2018, pp. 10–14. Finally, elements of Chapters 1, 8, and 9 appeared in *Environmental Humanities* as "Being Dumped," *Environmental Humanities*, 11(1), May 2019, pp. 180–193. All these texts are reproduced with the permission of copyright holders.

The photographs of the series *Carbon Black_Cruz Quebrada* were realized by means of a protocol developed during Anaïs Tondeur's artist residency at the European Commission Research Centre (JRC) in Ispra with the involvement of climate modelers Rita van Dingenen & Jean-Philippe Putaud.

γράψον οὖν ἃ εἶδες
καὶ ἃ εἰσὶν
καὶ ἃ μέλλει γίνεσθαι μετὰ ταῦτα.[1]

preface: dumped

Every day, scientific studies, media reports, and visceral experiences of the rapidly deteriorating state of the environment hit us with a growing and disconcerting force. In drinking water, microplastics abound, and, by 2050, the total mass of synthetic, human-made materials in the oceans is predicted to surpass that of fish biomass. Megalopolises on different continents languish under a stew of airborne toxins during the intensifying and protracted periods of extreme smog. Forest fires consume large swathes of wooded land, due to a combination of rising global temperatures, droughts, monoculture plantations, and meager investments into (as well as the unwillingness to rely on local knowledges for) fire prevention. Topsoil degradation, threatening the health and fertility of the earth, entails acidification, sharp increases in salinity, and toxicity, coupled with diminishing nutrient capacity and oxygen availability to plant roots.

Preoccupying as these raw empirical trends are in their own right, they are also indicative of a subtler alteration in the delicate conditions that have been up until now sustaining life on the planet. Water, air, earth and even fire (the four classical elements that, though they admit further additions, are shared by disparate philosophical and mythic

---

[1] "Write, therefore, that which you see, and that which is, and that which will become after this" (Revelation 1:19).

traditions) no longer correspond to our mental representations of what they are.[1] The image of water that automatically forms in the mind of a person hearing the word rarely includes plastic debris, cadmium, mercury, lead, coliform bacteria, and petroleum hydrocarbons. Thinking of air, we do not usually associate it with sulfur dioxide, nitrogen oxides, and particulate matter from forest fires or coal-powered factories. The meaning of soil does not tend to encompass heavy metals, phosphates, inorganic acids, pesticides and nitrates, polynuclear aromatic hydrocarbons, polychlorinated biphenyls, chlorinated aromatic compounds, detergents, and radionuolides. Whereas some of the elemental changes are visible (for example, those manifest in photochemical smog), a vast majority elude our senses and do not figure in the sphere of cognition.

The balance between the rule and the exception has tipped. Compared to the not-so-distant past, when the worry was about geographically circumscribed pockets of pollution, the regular environmental conditions nowadays are such that "clean" air, soil, and water deviate from the norm. We are yet to catch up with the strange reality the unintended cumulative consequences of our technologies and economies have spawned. Drawing near the current condition of water and other elements in thinking would satisfy infinitely more than the demand for accuracy that would culminate in an adequate representation of the altered object and a mental adjustment on the part of the subject. With such a sobering adjustment, we would also do justice to the rapidly vanishing, if not already vanished, world.

Though philosophy begins in wonder, it may end in dread. When profound enough, both of these affective states shake whomever is in them to the core. Contrary to the complacent perspective on the world according to the prefabricated structures of understanding, philosophy at its most radical is an encounter with existence, which comes to pass in an atmosphere of an acutely felt dearth of understanding, as if one has not experienced that which is so encountered before. It is this feature that immunizes philosophers (I mean philosophy not as a profession but as a vocation, a calling, a dedication, a way of life even) to a kind of jadedness, to familiarity with one's surroundings, seemingly undeserving of as little as a side glance. It is also this quality that imbues the philosophical attitude with a child's joy and curiosity or, on the opposite end of the emotional spectrum, with fear and caution in the face of the unknown. Among possible reactions to the latest transformations in the environmental elements and conditions, a benumbed business-as-usual approach, fostered by governments, corporations, and dominant ideologies, is not a viable option. Philosophy's flair for arranging an unparalleled rendezvous with the world is indispensable today, because we *are* confronted with a largely unknown world, sculpted by the lingering effects of industrial activity, for the first time in the twenty-first century.

In *Dump Philosophy*,[2] I tap into the discipline's unique strength of switching on an unorthodox vision of reality and propose to take stock of what the earth—the earthly fold synecdochically combining the rest of the natural elements—is, what it has become. I begin with the hypothesis that our planet is at an advanced stage of being converted into a dump for industrial output and its by-products, not to mention for consumerism and its excesses. The release of huge volumes of carbon dioxide into the atmosphere and the indiscriminate use of plastic bottles, bags, fishing nets, food wrappers and containers now ubiquitous in marine ecosystems are sufficient to qualify as *dumping*. Separate from these practices in space and time, their residues are no longer insubstantial traces in the air or in the water, but forces that reshape habitats, climates, and elemental milieus.

Assuming that we are not placated by clichés about the embeddedness of our lives and bodies in the environmental context, with which we are mutually constituted, we will be quick to discover that the environment's becoming-dump bears directly on our existence. Our diets, sensory possibilities, and statistically prevalent diseases (cancer, cardiac and respiratory ailments, diabetes, and so forth) are under such a powerful sway of the elemental mutation that corporeality, the physical or physiological fact of embodiment, is implicated in the workings of the dump. If we further subscribe to the view of the mind integrated into, rather than split from, the body, then we will see that the vicissitudes of corporeality have a profound influence on our ways of thinking. Ideas boil down to soundbites and buzzwords arranged in chains of free association; the flood of information submerges perception and cognition alike. The mind is no less affected than the body by the elemental mutation it has helped instigate. The dump penetrates the very fibers of our being, the processes and events that make us who we are: our humanity, animality, and vegetality, our

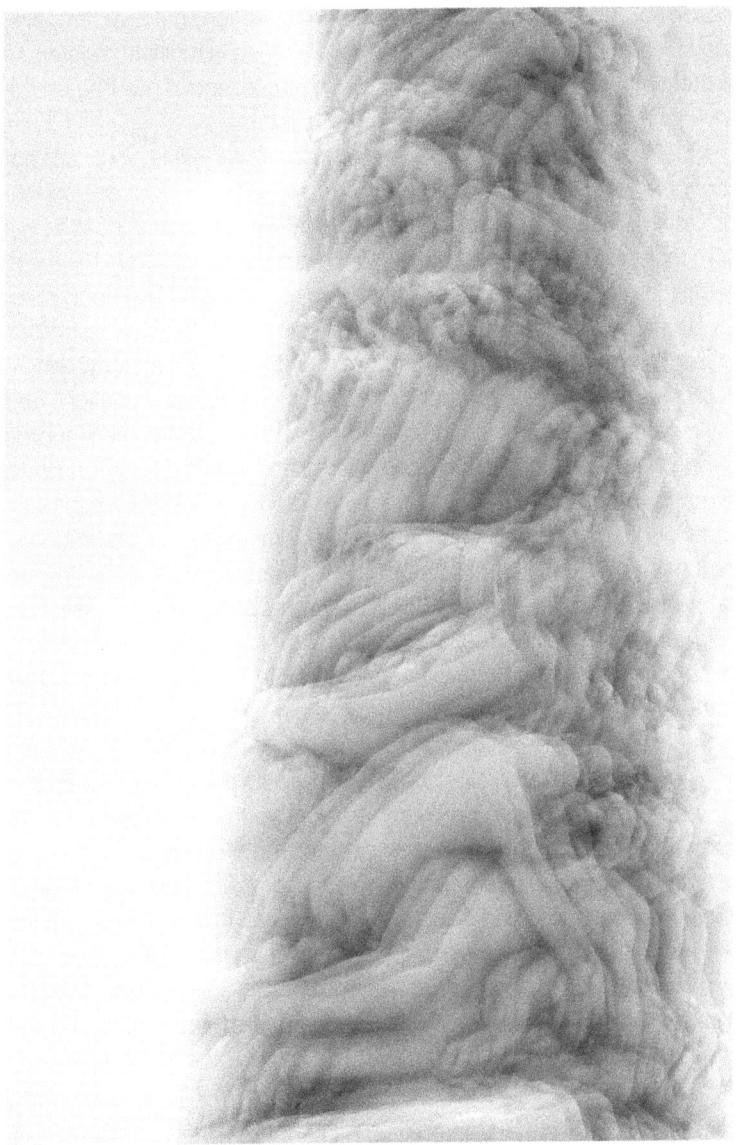

Figure 1 Drowning in, Anaïs Tondeur, 2018–20, Pigment print on Murakumo
paper, 42 × 63 cm

reasoning and organicity, sensation and perception, nutritive, emotive, and discerning capacities. Becoming entrenched in multiple registers of existence, it scrambles them, reproducing the effects it has had on the environmental elements.

Much of the book you are about to read seeks out an apt phenomenological description of the dump as the crux of our epoch. Or, rather, descriptions, taking into account the multiple senses of the term that range from the economic to the computational, from the act of defecation to a breakup, from a landfill to unbearable living conditions. Given the scope of the undertaking, you are invited on what promises to be a breath-taking ride through contemporary ontology, where fragments of thoughts and bodily functions, interpersonal relations and environmental factors, technological inventions and the mechanisms of psychological identification commingle and accrete in an uncanny, perhaps sublime, and definitely incoherent aggregate. We will pay close attention to the dump's theological and metaphysical forerunners, too: the religious and philosophical love-hate (heavily tilted toward hate) relation to unruly heaps, disorderly piles, and chaotic pastiches. Relevant historical lineages do not, however, diminish the shocking and, frankly, unprecedented realization of the most extravagant past fantasies and fears in our current predicament.

In order to begin the journey toward reclaiming the ability to act not only *within* the dump but also *on* it, we will have to evaluate the full extent, to which we are bogged down in the toxic pile that claims us for itself, body and soul. Onerous as it may be, the next step will be coming up with the *strategies of undumping*: uncluttering, revitalizing physiological, cognitive, ecological, and planetary metabolisms, reactivating becoming beyond mutations provoked by the dreams of immutability at every one of these levels. To use Hannah Arendt's expression, we will need to think without banisters, in the absence of tried-and-tested support structures for thinking in action. We will not have the luxury of leaning on catchphrases (renewable energy, geoengineering, entanglement, the philosophical leveling of hierarchies, etc.) because, in one way or another, they participate in the logistics of the dump. We will be groping for answers and, even more so, for the right questions in semidarkness, at the dusk of thinking and being.

The owl of Minerva, whom Hegel credited with the dialectical vision of the twilight, and the crane, who is one of Lady Yun-hua's (also

known as Yao-chi, Jasper Lady) apparitions, are taking flight from the
West and from the East, soaring over scenes of devastation, many of
them inaccessible to the naked eye, blanketed by smog, or otherwise
obscure. Under their wings—a planet ecologically and ontologically
mutilated, a dump for the mounting and non-metabolizable waste of
human industries and a desert encroaching on biological, cultural,
linguistic, and ideational diversity. In the dimming and desolate light
marking off their silhouettes, devastation itself recedes from our senses
and thoughts. But how does the disaster appear (in the afterglow of
which pyrodump?) through the eyes of the owl and the crane, gliding
on the flows of the aerodump above the vast expanses of the geo-
and hydrodumps? At the frontiers of imagination, *Dump Philosophy* will
survey extant being from the bird's-eye view of Minerva's owl and Lady
Yun-hua's crane.

globality

Ours is the age of the global dump. And the information without form that sometimes lends a name to postindustrial societies, economies, and ways of crafting knowledge is but a brushstroke in its portrait, the still or already unframed world-picture of our actuality.

We live and die in a dump of ideas, bodies, dreams, materials, snippets of relations, soundbites and memes, decontextualized and dehistoricized, produced as waste, reproduced ad nauseam, clipped, isolated and thrown together in a massive jumble in the wake of a world. What do the words *we live* comprehend?[1] How to apprehend them without arresting their referent? According to the preeminent ancient sensibility, they mean "we animate and are animated, move and are moved";[2] in the modern paradigm, they are likely to convey that we produce and reproduce (ourselves). Living in a dump, we are moved, produced, and reproduced by the dump, as by ourselves. For the most part and despite being alive in the medical sense of the term, we are dying there, dismembered, thrown out, trashed, alienated from our alienation, coming to love it or altogether indifferent, apathetic, no longer involved, anesthetized with pharmaceutically and ideologically manufactured painkillers. The dump lives us, lives *for* us. It takes over the movement, production, and reproduction of world-destruction, wrecking the very being-world of the world. One might say that the dump is unloaded on the world's frame—which it disjoints—more than on what is framed as worldly.

Metaphysical, religious, and moralizing proselytizers, living and dead, scream in our ears that we must wake up from the nightmare of our individual and collective lives while it is not too late, in time for repentance and conversion. They urge us to open the eyes of the mind or of the soul and finally to begin living, even if we are already in the concluding phases of our biological lives, abiding for the first time with truth or with god. As we shall see with yet another set of eyes, on the brink of every sort of vision turning inutile, the dump that lives us and lives for us is the coveted "true life" realized. To be precise, the dump is that life's unforeseen side effect, the result of persistently devaluing and trashing existence here-below, of treating the world as a vast wastebasket or, at best, as a mere springboard for the noblest, luminous, ideal, eternal being.

In the middle of a terrible nightmare, we wake up to a worse nightmare, falling deeper into troubled sleep. (Is it possible not to fall but to be dumped into sleep? If so, this is what's happening to us, not least thanks to our exhaustion, general sleep deprivation, and a growing reliance on sleeping pills among other pharmaceutical or biochemical aids.) The dismantling of old metaphysics has been declared complete. Yet, the work of disassembling its scaffolding and edifice is not a demolition derby: one cannot accomplish such a task once and for all. A brief pause is fertile grounds for resurrecting the tired, frayed, tattered instantiations of the metaphysical project that unabashedly claim to be new.

To add fuel to the fire, the work of mourning metaphysics, ongoing since the nineteenth century, has been not just paused but brusquely terminated. In exchange for metaphysics *and* for mourning it, we endure a narcissistic reopening of the wound in a melancholia that, beyond this or that human subject or group, is afflicting the world as a globalized whole. The business of the Anthropocene is one symptom of this malaise, this melancholic navel-wound gazing. Another is the reconstruction of ontology after metaphysics that culminates in being-as-residue. Here, being is leftovers,[3] morsels that fell from the table of nothing. Following the thread of both symptoms, the dump is an outgrowth of nihilism in all its positive splendor. Give the floor to Nietzsche's Zarathustra: "The desert grows: woe to the one who harbours deserts! [*Die Wüste wächst: weh Dem, der Wüsten birgt!*]"[4]

The global dump is a desert extending on land and in the hypoxic zones of the oceans. The more of it there is, the more it grows— mimicking the activity of what the Greeks called *phusis* and the Romans knew as *natura*—, the fewer the opportunities for future flourishing and finite growth. The vastness of devastation is at once vacant and full, spacious beyond measure and running out of room, barren and strewn with debris, a desert and a dump. Devastation de-vastates itself: we are aimlessly traversing the hyphen between the prefix *de-* and the vastness it at once negates and affirms.[5] Many species will not make it across this line, as short syntactically as historically, if grafted onto deep evolutionary time. It is uncertain that humanity will, either. With the desertion of being, the desert grows outside and within those who harbor it. We are deserted by being to the extent that we desert being. Today (better: tonight, in the creeping boundless night of the world), in today's tonight, *being is being dumped*.

Perhaps, a poisonous flower of nihilism, the desert blossoms from the inside, irradiating outward. Or, perhaps, the desert we harbor within arrives to us from the outside, searing with its dry heat every one of our thoughts, aspirations, retinal cells and intestinal tissues, the bronchial tubes and the lungs. In the outdated quarrel of materialism and idealism, it mattered where growth had commenced: in being or in consciousness, actuality or idea. None of this is significant any longer. The expanding desert is outside in and inside out, in the middle where existence used to be lodged.

It is not that the dump is over there, at a safe distance from the well-off members of affluent societies, who dwell at several removes from polluted water sources and open-air landfills. Radioactive fallouts know no national boundaries. Microplastics are as ubiquitous in tap and bottled water as mercury is in wild-caught fish. Smog does not stop at the municipal borders dividing the city's poor neighborhoods from the rich. The toxicity of the air, the clouds, the rain and the snow; of the oceans and their diminishing fish and crustacean populations; of chemically fertilized soil and the fruit it bears—this pervasive and multifarious elemental toxicity is also in us. At the physiological level, the outside slips in when we inhale and ingest it, the body's "hollow" interiors—the lungs and the stomach—exposed to the atmosphere, water, and food. But a philosophical explanation for this primordial infiltration is, I think, no less persuasive.

In keeping with a line of ancient reasoning, the body and its senses are microcosms that set apart, for the time being and in varying proportions, a tiny fraction of immense elemental regions: the heat of fire and its luminosity in the heart and the eye, the earth in the bones and the joints, water in the vital fluids. The elements are not the fundamental particles from which, brick by brick, cell by cell, molecule by molecule, we are cobbled together. The elements are not in us, or, if they are, only secondarily so. It is we who are in the elements as their proportionate and temporary circumscriptions. When proportions are out of whack, the imbalance restitutes a bulk of the delimited to exteriority. When the outside regions themselves are deranged and contaminated, so are their bounded segments. Toxic elements toxic bodies and senses make. And, since the mind is embodied, the list remains partial without toxic thoughts, desires, fantasies, and modes of reasoning that have, to be sure, also occasioned the evisceration of the world. With the acceleration of a positive feedback loop between exteriority and the psychophysical interiority that sets a bit of the outside world apart, their contents do not filter, ooze, seep, or percolate into one another. They are massively discharged, mutually dumped, instead.

Involving huge quantities of data and construction debris, the stuff of junkyards and a unilaterally declared end of an intimate relationship, excrements and a snapshot of a computer program's working memory at a given time, the flooding of foreign markets with extraordinarily cheap products and dreary living conditions, the dump is both outside and within. It relinquishes distinctions in physical space and the pivotal metaphysical opposition between the inner and the outer. Through its global reach, the dump swallows up and spits out *what is* together with the beyond of being, to which it was possible to elope as recently as the second half of the past century.[6] Its impact disorients and unsettles; it renders useless the habitual signposts for navigating complex, wrinkled, rippled, emplaced space. Tailored to the dump's uncanny measure, the world becomes a gigantic projection of the cave where Antigone was imprisoned by Creon, rather than an iteration of Plato's cave. There— that is, here—the death sentences meted out to all sentient beings living in the dump hinge as much on the deprivation and limitation of access to the basics of life as on the prisoners' exposure to a poisoned elemental "exteriority." And then the thought strikes us: there is nothing outside the cave (or, more exactly, there is no outside-cave: *il n'y a pas hors-caverne*), because life-giving elements are by now a hotchpotch of toxicity.

Conceptually speaking, the global dump is an achievement, indicating the way in which the much-maligned subject/object dualism has been overcome. Crude differentiation may be resolved either into finer differences or into indifference and undifferentiation. The crudeness of the subject/object relation is now replaced with the disorderly collection of –jects, often paleonymically called *objects*, and the haywire movement of –jection, oblivious to questions regarding points of departure and destinations. Late postmodernity has exchanged one of modernity's most important, if faulty, distinctions for an amorphous heap, which is not at all unheard-of in mythology and in the history of philosophy.

Not so innocent either, ecological, environmentally friendly, "green" discourses are implicated in the growth of the desert and the dump they abhor. As they rave about the butterfly effect adopted from a key figure in chaos theory, Edward Lorenz,[7] and aver that everything is interconnected, ecologists destroy much more than the category of causality and predictability with its illusion of control; they harm the fragile logic of articulation, the prelogical arc of *logos*, and the precondition for establishing relations. The moment everything is linkable to everything else with the same intensity of association, nothing is related to anything. Relations are stitched together of varying energies, degrees of exclusivity, the push-and-pull of the inbetween. In a word, of differences. It follows that undifferentiation combined with indifference is lethal to relations.[8]

Starting from the mental act of paying attention that singles out, is provoked or convoked by, and relates to a *this*, consciousness is partiality and discrimination, selective adherence and devotion. It neither predates nor survives its unique attachments.[9] The unconscious, as well, consists of multiple cathexes, the irregular investments of libidinal energy into an object. But the impersonal consciousness that predominates in the dump is a consciousness torn out of its relational dynamics, uncathected, and dumped, unable to rise even to the level of the unconscious.

In existence where everything is interconnected, everything plummets haphazardly into the same heap. It all ends up on a global dump, which englobes us on the outside and clutters us with its desert emptiness from within. The cognitive state suited for this condition is the unmitigated distraction that tears to shreds the ties of consciousness to that of which it is in each case conscious. Dumping someone after a period of infatuation does not just terminate a relation; the act disposes of relationality. As does fusion with the other. Trendy entanglements barring a modicum of *dis*entanglement contribute to the dense mess of dumped being. Although they proceed from opposite directions, the act of cutting relational bonds and the *pathos* of suffocating in their indiscernible proximity converge. Resigned in the face of the nascent dump, Heidegger had a premonition of its global approach: "Unavoidable is a confused *entanglement* [*die wirre* Verstrickung] in the massiveness, the boundlessness, the hastiness of the present at hand."[10]

all the world's a dump

What if we were to stage—assuming that staged it can be—the global dump with Shakespeare? I am thinking of Jacques' monologue in *As You Like It* and its feted phrase "All the world's a stage" (2.7.140). Less widely known is the replica Duke Senior makes in advance of Jacques' speech: "Thou seest we are not all alone unhappy. / This wide and universal theatre / Presents more woeful pageants than the scene / Wherein we play in" (2.7.137–9). Duke Senior's observation transports the spectators from Shakespeare's play and from the Globe Theatre to the theatre of the globe, "wide and universal." At this juncture, the dramatic device of a play within a play reverses the insertion of a miniature performance into the main action on stage by encrusting scenes from *As You Like It* into the world's still "more woeful pageants." At any rate, the Baroque strategy of a play within a play endows the drama with infinite depth[1] in the Hegelian sense of bad infinity (lacking closure, the series of X within X—here, a play within a play—may be continued indefinitely) *and* of good infinity (the speculative reflection of X by its double: the play on stage is a mirror for the play of the world, which, in turn, mirrors the one on stage).

As a stage, the world is a place for the appearing of what appears, for the phenomenality of phenomena. If *all* the world is a stage, then it is the place for the appearing of *everything* that appears. To ask what a stage is would be to ask about the meaning of the world. In the first place (for this is, indeed, a matter of place), a stage is an open, flat, and hard surface propping up the action, *that upon which* whatever or whoever appears makes the appearance. With the addition of backstage, crossover, and wings, it marries the visible to the invisible, the lights and the shadows, the appearance itself and the infrastructure for an event that, while not appearing, prepares its arrival. A world stage, the world *as* stage, is the theatricalized determinacy of an opening, an aperture on the horizon's surrounding closure.

Should we replace the stage with a dump, and do so not on a whim but obeying the strict "apocalyptic" injunction to write down that which we see around us,[2] we would immediately notice that the firm support the stage has hitherto provided is withdrawn from us. Unfathomable depth is essential to the dump; an underlying surface belongs to the stage. A dump is *that into which* beings fall, the dynamics of falling, and the state of having fallen. It is a nonplace for the disappearance of the dumped entities, and, assuming that *all* the world is a dump, it stands for that in which everything disappears in the very instant of coming into the world. Evanescing with dumped beings is the world stage, which is never empty. The dump is the unworlding and the unstaging of the world.

"All the world's a dump" declares, therefore, that all the world's been lost. The verticality of the fall supersedes the horizontal surface of appearance on stage, making performances of a play within a play unnecessary: a dump is always a dump within a dump to the power of infinity. It supplants the society of the spectacle with concealment behind a barrage of cast ideas, sensations, part-objects, bits of information, spent packaging, obsolescent artifacts, bodies living and dead, toxins, raw sewage. The dump's depth is not an effect of objective arrows pointing up and down that organize its volume. Its depth is conditioned by falling, dropping, discharging, and jettisoning, its vertical axis the sum total of these activities. The massification of the fall deepens the dump and finally knocks the bottom out. Technically, the dump is an abyss.

The time to save this or that being from dumping has run out. We have no more time to save being and saying from and through time and forgetting, as Plato set out to do in *The Republic*.[3] *This* being—any *this* including this *I* who is writing these lines, these written lines themselves, and you who are reading them—appears not as disappear*ing*, not as finite, but as, from the outset, disappear*ed*, finished, done for. Concealment also varies historically: at present, it is due to a layering without any particular order, through the erratic agglomeration of falling stuff that does not spare us, that animates and produces us, that lives us and leads us to death.[4] Despite the word of The Apocalypse (of Saint John) and also despite the word *apocalypse*, the end of the world

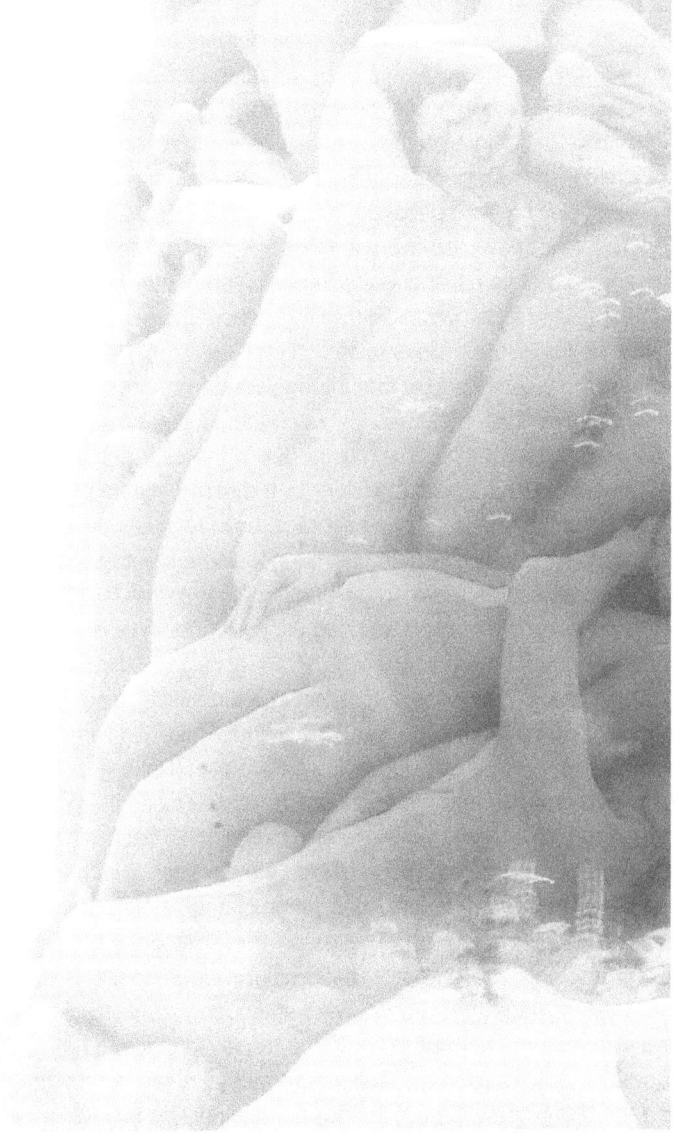

Figure 2 Drowning in, Anaïs Tondeur, 2018–20, Pigment print on Murakumo paper, 42 × 63 cm

cannot be staged after the curtain has risen on the final act. A nonevent, it occurs offstage, not in the world, imperceptibly, obscenely. Woeful as they may be, the pageants of universal theatre are not (yet) the end of the world. There can only be *signs* of the apocalypse, that is to say, indicators to be interpreted, traces, present absences and absent presences—never appearances.

The society of the spectacle kept an interval between the spectator and the talk-show[5] of everydayness. Mindful of the distance their functioning requires, Hegel classifies vision and hearing as ideal senses on the heels of a philosophical tradition, where the ocular relation was a prototype for Platonic ideas (*eídō* = "I see") and for Aristotelian *theōria*, "divine vision." A critique of the privilege Western thought has accorded to ocularity replicates the cleft between the subject and the object. But must we bridge the invisible gap between the seer and the seen, the gap opening up the field of vision, in the same style as that abstract cleft? Does the society of the spectacle have to give way to a dump, a thick cloud of debris shrouding us, filling the space between the eye and the visual object? Is there no fall-back solution to the fall of wreckages upon wreckages that blind and gag us, surrounding us on all sides and occupying us from within?

The gaze, which right before being extinguished sees that all the world's a dump, is that of Paul Klee's and Walter Benjamin's Angelus Novus. "[A] storm is blowing in from Paradise; it has got in his wings with such violence that the angel can no longer close them. This storm irresistibly propels him into the future, to which his back is turned, while the heap of ruins before him grows to the sky [*der Trümmerhaufen vor ihm zum Himmel wächst*]."[6] An indiscriminately quoted titbit of cultural theory, this fragment from "Theses on the Philosophy of History" has plummeted into the dump of ideas.[7] But it does speak of the dump, the heap of ruins that grows, *wächst*, much like Nietzsche's desert, and vicariously puts itself to work or into play in place of nature and its vegetal activity of growing. The discrepancy is only in *how* the desert and the dump grow: silently increasing the scale and the dimensions of nihilism's eerie plant or violently and suddenly blowing in with a storm. And the story does not end here: a bifurcation in the *how* also applies to scenarios that would see the conditions for life on our planet damaged beyond repair, whether gradually through global warming or abruptly through a nuclear holocaust.[8]

It is said that a weather system dumps precipitation when rainfall or snowfall is copious. The debris storm of world history does something else: it mixes the above and the below, causing the sky to fall on earth and the earth to be swept skywards in the heaps of ruins progress has wrought. The dump is an abysmal stage devoid of a lower horizontal support. It upstages space, triggering the collision and collapse of spatial extremes: the high and the low, left and right, front and back. Stage directions do not apply. In the tornado of world-historical debris, it is debatable if we, alongside everything that and everyone who is dumped, are falling downward, drawn by the force of gravity to the earth, or upward to the sky. Because the storm has thrown the sky onto the earth and the earth into the sky, the confusion is not just in our heads but in the dislocated elemental regions themselves.

So, all the world's a dump. When it was a stage, "all the men and women" were "merely players," with "their exits and entrances" (2.7.141–2). In a world-dump that is, for all intents and purposes, a nonworld or an unworlded reality, mountains of detritus barricade the exits and players do not make their entrances (they are thrown into the maelstrom). This thrust gathering momentum, the turbulence of dumping, substitutes the disquiet of existence. Being-in-the-world is being shaken in a bottomless wastebasket, being wasted, falling to no end, endlessly, and utterly unprotected from the deployment of such endlessness for sinister and highly specific economic or political ends. Thrown into the world-dump, existence is thrown out. In is out, outside-in. At this stage of humanity (in the temporal, rather than the spatial drift of staging) the meaning of being is unveiled with an exaggerated, grotesque panache as what being has always struggled against: a dump.

For Shakespeare's Jacques, the stages of humanity structure life's play. There are seven of them, sorting existence according to discrete rhythms and patterns that, for all their heterogeneity, rotate in a circle, the end a black-and-white photocopy of the beginning: "Last scene of all, / That ends this strange eventful history, / Is second childishness and mere oblivion, / *Sans* teeth, *sans* eyes, *sans* taste, *sans* everything" (2.7.164–7). Where Shakespeare ends, Beckett begins, his characters trapped in the oblivion of posthumous existence. We are contemporaneous with the tragicomedy of everything "*sans* everything": our experience—if the word is still apposite, which is doubtful in the extreme—is the "second childishness" of humanity grown so old that it unwittingly slips back into its infancy. The stage of no stages, when the fine-grained textures of time and space are sanded into an amorphous mess, the age of the global dump is the age of all the ages and *sans* age, ageless. Is that our impression of eternity, the "last scene of all" with props made of things that will not age, wrapped in plastic, swathed in carbon emissions, and stocked with spent nuclear fuel rods?

Jacques' "strange eventful history" teemed with events. Its rhythms kept their differences and, in a string of typical moments, became stages and told a story. The dump wipes out narratives and events, stages and sequences. Absent the events in it, the dump itself is the event of dumping into and out of existence, into *as* out of existence, in a paradox twenty-first-century existentialism (or whatever remains of it) cannot afford to ignore. It is an event threatening to cut time short, to undercut the temporality of time, that is. Without the senses (vision, taste . . .) and without sense, without the elemental microcosm and cosmic elements, not in a prelude to sense-bestowal but in the muteness of "mere oblivion."

The totality of the world Shakespeare stages encompasses body parts and the perceptual apparatus, institutions like the school (2.7.148) or the military (2.7.150), desire, and the emotions of hope and despair. His world is a shared milieu, a mental theatre of inner representations, and a body with or without organs on stage. Upstaged at the world's end, psychic, physical, social, and aesthetic theatres disintegrate, their walls crumbling. The psyche is not reunited with but dumped into the body at the junction of neural networks, biochemical cues, algorithmic functions, and circuits of information-processing. Social space is poured wholesale into the psychophysical dump via the channel of new communication technologies. Whatever remains of nature reaches our stomachs full of pesticides, antibiotics, or both, and it strikes our senses not before being mediated by *National Geographic* documentaries, *Animal Planet* shows, and *Nature Sounds* apps. To the alternatives of segregation and integration one is obliged to add dumping, which desists from articulating beings among themselves as much as from maintaining them apart.

mechanics

the fall, massiveness, piling up

The rudimentary features of the dump are already tangible, if not apparent. Comply with the basic laws of physics as they may, these features secretly bear allegiance to metaphysics. More than that, they are not the static descriptions of a phenomenon but clues to its operations, to the modes of the dump's activation, the manners of putting it to work. What a dump is depends almost entirely on what it does. To study its activity is to risk lifting the lid on the prime mover of our age.

1. *The fall.* Dumping is causing to fall. In Old Norse, *dumpe* means "to fall suddenly to the ground or into water," to plump.[1] Onomatopoeic, the word falls on the ear with the explosive or, more accurately, the plosive sound of the consonants *d*, detonated on the tongue blade, and *p*, flying off the lips. It begins and ends by instantaneously stopping the airflow and blocking the vocal tract. Which is to say that it begins and ends in suffocation. Of speech, above all. (Is that why one can only bear witness to it in writing?) As soon as we pronounce it, the dump takes our breath away and robs us of words, the voice, articulation, the bread-and-butter of philosophers enamored of *logos*. Suspending breathing that, over and above an exchange of gases between an organism and the atmosphere, signifies life and embodied spirit, the dump actively instigates the very fall, the culmination of which it substantively marks. It reunites falling and fallenness, with the difference between the two transposed onto the verb *to dump* and the noun *a dump*.

In the former, verbal modality, the unanticipated and violent force inherent to this act favors the connotations of *making fall* over *letting fall*. Instead of delivering beings to their fate, dumping hastens their fall, speeds up their ruination, quickly drops them into nonbeing. It does so by way of getting rid of them, throwing them out, making them fall away. Away from what or from whom? We will circle back to this question, behind which a venerable theological tradition frames sin within the dynamics of the fall, enhanced with the Gnostic derision of the body.[2] The entire history of metaphysics is that of spirit trashing matter, dumping matter, and thinking that matter is a dump. Actually, matter is only a dump because it has been dumped and its fall precipitates spirit as that from which the fallen has fallen. Spirit is a sublime dump truck, tirelessly offloading beings into nonbeing, emptying being out en masse, and transforming the metaphysical equivalent of gravitational-potential into kinetic energy in the process. It comes into its own and coincides with its concept when it finally contains nothing, having released its great accumulated mass. But it also comes to its demise: carrying no matter, the dump truck of spirit buckles under its weightlessness, as it turns out that the volume of its gigantic bed had been generated by the load it dropped.

2. *Massiveness.* The dumped falls both suddenly and massively, in great quantities and in the shapeless shape of mass. The precipitous and relentless nature of the fall means that there is no time to sift through it, to absorb or to adjust, to make room. Dumped matter is heaviness, a quality automatically transcribed into a measurable quantity, that pulls existence down to earth, except if elemental confusion prompts us to misidentify the trajectory of the fall. Mass is matter in bulk and unbearably light, a dump and a desert, the oppressive gravity of weight and the arid ideality of number. You will not recognize the outlines of material entities in it. The mass of a massive throw is not that of an object; it is mass as such. In the gargantuan outpouring of dumped existence, the principle of deindividuation is beyond dispute. Whatever entity, the

Scholastic *quodlibet ens*, falls alongside countless others, human or not, and is indistinguishable from (precisely in the capacity of the generic *whatever*) less-than-an-entity. The dumped one is immediately the many, not in a company with them but as a consequence of their and "our" massification. The sinner's fallen body is sentient and anonymous mass, fleshy matter denuded of form, a *memento mori* flashing in its nonindividuation a snapshot of the corpse it will have been. Judeo-Christian flesh must submit to the grueling routines of disciplining, punishment, and mortification to obtain a semblance of form. With the soul and its formative influence out of the secular picture, flesh follows suit and vanishes. Living corporeality mutates into the undifferentiated agglomeration of biomass, expendable organic matter burned to derive energy, the materialist stand-in for spirit. For, the massiveness of dumping *is* its energy: the bigger the pyre that sends incinerated matter up into the sky and the harder the rest drops down to earth, the more force such discharge, impact, and combustion provide. A heap of shapeless stuff that stays behind, the dump in the state of rest, an "ill-shapen piece" (as per the word's other plausible etymological lineage[3]), is a trace of those past activities.

3. *Piling up.* The dump that the earth, the sky, and everything inbetween have become is a peculiar, internally disarticulated, disorganized, discombobulated totality. Sweeping all that is up into their midst, the sprawling piles of stuff do not articulate the fragments of bodies, ideas, materials, and information they contain. These adjoin one another at random, with no rhyme or reason, with no *logos*, with no *with*, commemorating the moment when, by fluke, they have been thrown out together. A comprehensive definition of the dump would thus be *a massive, undifferentiated, and indifferent fall*, whence "our" age receives *its* identity and *its* difference. That's how nihilism, self-assured and productive, takes on a body, how it gives a body to itself, replete with the imitation of life's essential movements. We may applaud the promise of freedom in the dump's disorder, welcoming an opportunity for liberation

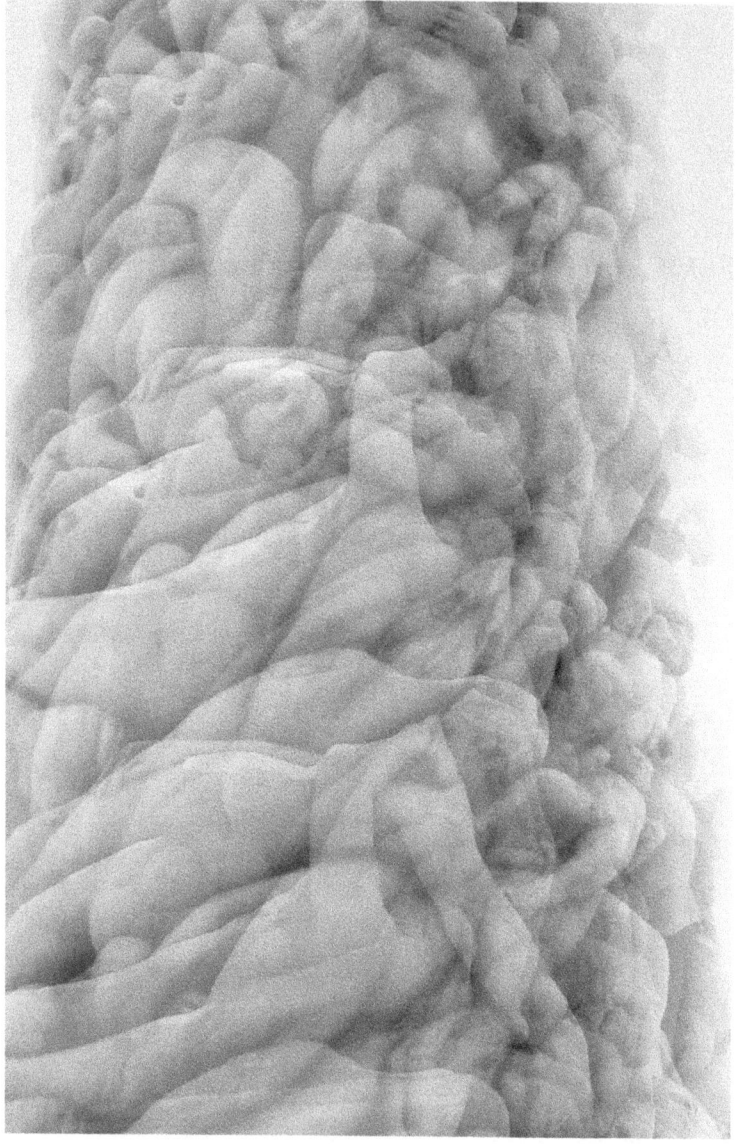

Figure 3 Drowning in, Anaïs Tondeur, 2018–20, Pigment print on Murakumo paper, 42 × 63 cm

from the stiffness of teleology and a refreshing anarchy of the mélange. True: the totality of the dump falls down and falls *apart*, staving off a fascist-organicist variety of totalitarianism. But its indifference is so thoroughgoing that it lies on the hither side of sameness and difference—yet another staple metaphysical dichotomy leveled. As its piles grow, they choke off existence and clog every modality of time, including the past. The dump's increase is unlike vegetal flourishing from the inside out, partially co-opted by the desert. Unless the debris is swept up by a storm, it grows by accretion, less regular yet than that of minerals. An actual nightmare stems from the daydream of metaphysics and of capitalist economy: monstrous growth without decay in a dissociation between two vectors of movement (*kinēsis*) emblematic of Aristotelian vegetality. Their uncoupling jams the third kind of movement, namely metamorphosis. The imitation of *phusis*, of the self-emergent growing whole, in economic and metaphysical growth is partial and shoddy. The pile piles on indefinitely, just as its components are ground down in a seemingly infinite quantitative diminution, reflecting the ontological poverty of modern growth restricted to a numerically mediated increase. As the maniac activity of dumping massifies the fall, the fallen bits for the most part do not rot and, not rotting, accumulate. The nastiest sectors of our dump do not stink and evade decomposition. They are made of stuff that refuses to be unmade and add on more of the same: polyethylene and depleted uranium, cryopreserved corpses and the ideals of immortality, life everlasting, immutable reality. In the red forest located in and around Chernobyl's 30-kilometer exclusion zone, the dump has been thrust upon plant growth and decay. There, pine trees turned reddish and withered shortly after the accident, their trunks gathering on the ground over the last decades. Piling up on the forest floor, they are not decaying as they should, nor being digested into the earth.[4] Beyond the reach of the senses, radiation has meddled with the vulnerable ecosystems of decomposers: fungi, microbes, insects. The death of death fatally injures life, the dumped leftovers from our mad aspirations to incorruptibility, to static preservation, interfering with the narrow parameters of vitality. We perish of our perversely realized yearnings for immortality and drag much of the biosphere down with us. Contrasted to the dump's petrifying power, which accords with sensory deprivation in Shakespeare's "last scene of all," the smell of rotting is the whiff of salvation.

Now that the dump's contours are within view, it's time to study the nitty-gritty details of its mechanics.[5]

falling before and after the death of god

Aristotle invented matter. As it happened with much of the philosophical vocabulary he developed, rather than create the concept out of thin air, he added another layer of meaning to the already existing everyday word for wood, *hulē*. The provenance of the philosopher's *matter* is a perfect illustration of the responsibility (*aition*, often translated as causality) assigned to it: matter is that out of which something arises, *to ex hou gignetai*; the arising—also of its concept—is not out of nothing.

In *Ennead II* "On Matter" (*Peri hulēs*), Plotinus is nonetheless adamant that matter's defining trait is "to be 'the underlying' [*hypokeimenon*] and 'the receptacle' [*hypodochē*]" (II.4.1.1–2). These are the respective signatures of Aristotle (*Physics* A9, 192a, 33) and Plato (*Timaeus* 49a, 6), who, for all the disagreements between them, share the supposition that nothing can drop beneath that which is *under* (*hypo-*), demarcating the lower conceptual fringes of vertical space. If that is matter, then, shifting the emphasis from the arising to the ground of arising, it refers to *that into which* one falls, returning to this ground. Or, as Plotinus puts it, "the depth [*bathos*] of each individual thing is matter: so all matter is dark [*skoteinē*]"[1] (II.4.5.9). To bring matter to the surface, to shed light on it—for example, as mass—is to dematerialize it. A sound approach to matter compels us to plunge into its depths, to go under, and the only safety nets capable, within limits, of foiling a nosedive are the visible shapes, "forms," images, *eidoi* (II.4.1.2) that the receptacle receives close to its superficies.

Ancient matter is evidently anything but a stage, upon which action appears in the limelight. Nor is it a dump. Whatever falls into matter gets embraced and enveloped in a tailor-made garment of material form appropriate to the being that falls, so much so that it is hardly distinguishable from that being's skin. The receptacle is adjusted to each entity it receives in a unique blend of truth and justice. Things fall in their places, the fall coeval with the creation of place. If at first they don't, if their places are not fleshed out, the receptacle will sway, vibrate, shiver, and shake so as to guide things to that which is proper to them (*Timaeus* 52e). Regardless of its usual associations with passivity, reception is an *act* of acceptance that, by exclusion, rejects much else. Stretching out from below, matter is not an inert buttressing structure; it is not an invariable foundation of existence, but a repository for preconscious discernment, actively receptive toward the given.

Aristotle declines to talk about the fall Plato mythologizes in his speculations on the receptacle's vibrations. An event of the kind will have been[2] immemorial, unrepresentable, unavailable to experience. The code word for Aristotle's silence on the subject is *hylomorphism*, stipulating that matter reaches us already teeming with forms resistant to further analysis. The fall rings in his ears with a much more literal sense of various entities gravitating to their predominant elements, their native regions. Russian philosopher Vladimir Bibikhin points out correctly that the Aristotelian earth is that unto which things tumble. But he goes astray in asserting that it is "a cosmic dump [*kosmicheskaya svalka*], whither everything falls down and gets mixed up."[3] (Let it be mentioned, in passing, that there is also the relatively recent phenomenon of space junk orbiting the earth without, for the most part, making a descent into the atmosphere. Since mid-twentieth century, our debris has been hurled so far that, in a suspension of its fall, our planet's vicinities, too, have become dumping grounds, or nongrounds. This milestone in the history of technology merits a significantly more detailed treatment than I can extend to it here. Suffice it to say that, besides the earth as a cosmic dump, humanity is steadily marching toward a transfiguration of cosmos into a dump—the very opposite of the shining order that is the Greek *kosmos*—in a tendency likely to accelerate should plans for *Homo sapiens sapiens* as an interplanetary species be concretized.) Bibikhin omits the fact that the centerpiece of Aristotle's *On the Heavens* where the earth is made intelligible within a cosmic frame is the centering of the fall and a formulation of laws that attribute to it a way of gravitating toward the center of the earth, itself situated in the center of the universe. The unequal velocities of falls have to do with adjustment and with material, hyletic justice: heavier objects are reunited with their element faster and more thoroughly than the lighter ones, with a lower proportion of earthy density in them. Does this pattern bear any resemblance to the cosmic dump, a pell-mell destination for a generic fall and its mixed-up repercussions?

In a leap from physics to a theologically inflected metaphysics, the earthiness of existence is responsible for its earthward plummet. The fall of Adam and Eve diverts them back to their native element, to the earth, *adamah*, out of which they were crafted (Gen. 3:19), divine punishment rehashing the precepts of hyletic justice. Neoplatonism and Gnosticism are unanimous in their take on matter as that which falls, or that which causes the fall, the body pulling the soul down. Physical existence not only taints but also adds a foreign weight, uncomfortable at best, to metaphysical being. Plotinus "seemed to be ashamed of being in the body" (Porphyry, *On the Life of Plotinus*, 1, 2) and extoled the virtues of the soul. Death for him was an instant of liberation from the alien mass of corporeality to be discarded as a distraction, a burden, deadweight, a corpse one carries in the course of a period one perceives as one's life rather than a faithful substratum that carries one, a clumsy piece of packaging that distorts its precious immaterial contents. In short, the body was the dump of the soul and, in this capacity, it had to be dumped.

It was not what the body did that sealed its fate but what it was: evil. With a theological twist, to fall into sin was to fall into the body, to confirm the brute fact—the facticity—of embodiment. (Today, the view of the body as dump persists on a molecular, genetic level, with noncoding DNA deemed to be "junk DNA," which amounts to more than 98 percent in the case of the human genome.) A barely tolerable body had to be purged of its base appetites, desires, the Augustinian *concupiscentia carnis* (lust), and even of its reliance on perception. Finally, it had to be purged of itself. Paul's words in 1 Corinthians 13:12, "For now we see through a glass, darkly [*en ainigmati*]," paint the picture of fallen vision that muddles more than it divulges reality, presenting the world in the form of an enigma, withholding the world (and, most of all, the truth that resides in the otherworldly domain) in the course of its very self-presentation. This fallen vision is the vision of the fall. Unawares, it inspects the dump where darkness and confusion predominate and where the talk-show of everyday life, the wheeling and dealing of zombie existence, revels in a series of *enigmas*, "dark riddles" or "obscure sayings." The fall, seen from the heights of spirit, engenders the darkness of material depth and awakens in Paul a nagging nostalgia for the surface, the countenance, the direct look of things (a version of the Platonic *eidos*) he rushes to meet face-to-face, *prosōpon pros prosōpon*. In a dump that in Paul's eyes is worldly, in the dump that *is* the world, relations (*pros ti*) are nonexistent, whether among a befuddling diversity of surfaces or between surface and depth. The vis-à-vis he is eagerly awaiting is a metonymy for relationality.

Following in Paul's footsteps, Augustine identifies in himself an inner dark vision, as he reminisces back to the time of his unmitigated fall. He confesses that, before his conversion, in his youth, he was hardly able to discriminate between the "fire of serene affection" and that of "libidinal desire": their mishmash "beclouded and obfuscated my heart [*obnubilabant atque obfuscabant cor meum*]" (*Confessions* II.2.2). His core, his heart, was a dump where divine love and human lust interflowed. The confessionary act sweeps through the psychic dump, picks and chooses, decides, and so negates its dumpiness. The confession cleanses inasmuch as it insulates, by means of a judgment that exceeds the scope of rationality, one kind of love from another. Augustine admits that his narrative recounts the "carnal corruptions of my soul [*carnales corruptiones animae meae*] not because I love them but because I love you, my god" (*Confessions* II.1.1). On a sympathetic reading, he practices the wisdom of love, as opposed to the philosophical love of wisdom. His is a discernment that resorts not to concepts and categories but to fire, the light and the heat of love and, indeed, of the love of divine love: "*amore amoris tui facio istue*," "for the love of your love I do this" (*Confessions* II.1). The wisdom of love is a remedy against the confusions and obfuscations of jumbled being that has all the trappings of nonbeing. Confessions undump the soul without giving up on the body.

Augustine further rescues matter, heaviness, and the fall through the power of love permeating creation. Inanimate objects partake of this power by virtue of their attraction to the places where they fall "for the weight of bodies is, as it were, their love [*Nam uelut amores corporum momenta sunt ponderum*]" (*The City of God* XI.28). Within a modern scientific framework, the claim is bizarre. But, to an Augustine-inspired mind, a science seeking the laws of gravity (or any laws for that matter) with a meticulously, methodologically cultivated aloofness is a forerunner of the global dump. Even then, this science lies to itself: it passionately loves the pure objectivity, presumably stripped of passion and personal attachment, that Francis Bacon's has built on the ruins of the smashed idols. By contrast, attraction and repulsion, love and hate, are conspicuously missing from the fall of the dumped, discharged with such insouciance that it is impossible to conclude with any degree of certainty whether they are falling up, down, or sideways.

Admitting love, Augustinian matter undergoes spiritualization. As he subtly retrieves Aristotle's *hulē*, Augustine infers in one of his sermons that wood is unsinkable and, thus, undumpable. Together with fire, or perhaps *as* fire, it is materiality verging on spiritual ideality. If the world is a sea of temptations and sins, then, while keeping in touch with dark and liquid material depths, the wood of the cross carries the believer on the surface of the waves to salvation: "So it's essential we should stay in the boat, that is, that we should be carried on wood, to be enabled to cross this sea. Now this wood, on which our feebleness is carried [*hoc autem lignum, quo infirmitas nostra portatur*], is the Lord's cross, with which we are stamped and reclaimed from submersion in this world."[4] The wood of the cross is matter so light yet sturdy that it floats and hoists its load, not letting it sink stone-like. (Augustine ascribes mineral heaviness and matter external to spirit to Judaic law, which he juxtaposes to Christian love, the "law of the heart": "It's stones, you see, who are throwing stones, rock-hard men who are stoning you. They [the Jews] received the law on stone, and they throw stones."[5]) Spiritual matter inhibits the fall of what is so shaky and unstable as to be always on the brink of falling—our feebleness, the infirmity (*infirmitas*) coextensive with the sentient body—and redeems what is already fallen—carnality, flesh as such. It is the underlying, though not rock bottom; the abyss of spiritless matter rages beneath it. Slotted between darkness and light, between depth and surface, Augustinian wood of the cross recaptures Plotinian matter of the mind, *noēte hulē* (II.4.5.40), the mediator that cushions the fall.[6]

Back to Plato and to the origins of Neoplatonism! The line that has exerted the greatest influence on Plotinus goes back to *Phaedrus*: "the soul of all cares for the soulless [*pasa hē psuchē pantos epimeleitai tou apsuchou*]" (246b). Not by accident, care and engagement are waning in the global dump, where all are soulless, deprived of animating principles, save for a certain vibrancy feigned in the agitation and turmoil of dumping. As for the phrase from *Phaedrus* concerning care, it occurs in the allegory of the chariot, exploring the "joint potentialities," or the "powers that grow together," *sumphutō dunamei*, of a human soul (246a). One of the two horses pulling the chariot of the human psyche frequently refuses to cooperate with the other and with the charioteer: the first, "fully winged" and consanguine with those of the gods, mounts up and passes on-high, *meteōroporei*; the other, having lost its wings, "is borne along until it gets a hold of something solid, when it settles down, assuming a body" (264c). With its weight and solidity, the body, which the soul appropriates, actually takes possession of its proprietor, whom it drags down to earth, *epi tēn gēn*. The animation of mortal living beings is a contest between their inner antagonistic potentialities to rise or to fall, the powers that, despite growing together—from the same root, one assumes—strive in opposite directions in the Platonic *agōn eschatos*, the struggle of ends (247b). This irresolvable tension is a source of energy for a finite life. Neoplatonists and Gnostics propose diffusing intrapsychic strife in favor of the perfectly winged portion eager to ascend; the dump eliminates the tension by opting for a free fall of living bodies recast as mass.

Heads or tails, Neoplatonism and the global dump are two sides of the same coin. In the twenty-first century, the view that material existence as a whole is garbage has triumphed. The concept of dignity and the apparatus of rights offer a Band-Aid solution in this predicament. Existence falls; it is a fall and fallenness, with massification as the only difference between its ancient and contemporary statuses. But, to step back to an earlier question, what is the *topos ouranios*, the ethereal site, the utopia of on-high, from which beings are offloaded? Particularly after the death of god and the waning faith in Spirit, Reason, and Progress as his other manifestations? What remains below, once the otherworldly *above* has been eliminated?

We are firmly, if baselessly, convinced that what remains below is the world and the earth, roughly corresponding to Augustine's spiritual matter and spiritless matter, respectively. Falling back unto them, we seem to complete a millennia-long odyssey and come home to everyday life and finite thinking, having sailed past the Scylla and Charybdis of metaphysics and back. But *the world, this world*, is a theological construct, the created realm intelligible exclusively with reference to its creator, or, more broadly, to the work of spirit. Moreover, the world is a stage and the play performed there, not the dark depths of matter divorced from form, which is under a permanent suspicion of having smuggled slices of idealism back into our thought. The earth, then? Because "There is no Planet B!"[7]? But the earth has long been god's dump, which after his death, is (again) our "own." And we can only return to it as to a dump, to be dumped into it. The geological epoch of the Anthropocene advertises this return, of the living and the dead, by illuminating the accumulated vestiges of human activity in a layer of the earth and confirming: we now inhabit a geodump. Reacting to anthropogenic climate change, geoengineering endeavors to impose a form disengaged from matter on the material dump of the earth. In the words of Plotinus, the project's unintended outcome will be a cosmetically embellished, cosmeticized, cosmicized corpse, *nekron kekosmēmenon* (II.4.5.19), an embalmed cadaver of the planet.

Dumped, excreted from dead divine intestines (Augustine configures memory as "the stomach of the mind" [*Confessions* X.14.21], and our remembrance in god is safekeeping in his infinite stomach for all eternity, a sort of divine indigestion) is existence shrunk to or distended into unformed matter. Thereafter, matter is abbreviated to mass, and mass is primed for the extraction of energy, be it in the guise of labor, be it as knowledge, calories, or time. Going down this road, it is possible to fall beneath the lowest threshold of the underlying, as Augustine foresaw in another context. The fall of mass-energy, of what matter is in up-to-date physics and outside its disciplinary confines, is without local adjustments and shorn of justice. If anything, it subsumes locality to itself, its admittedly uneven distributions generating space as an appendage to the possibilities of energy storage and extraction.

Having fallen, nothing and no one have their places in the dump, where the place for place is absent. In lieu of experiencing the anonymous welcome of the receptacle, they join the chaotic heaps of the dumped, crushing and being crushed, trampling upon, flattening each other. Is *that* the end of hierarchies? The struggle for survival, consistent with Social Darwinist script, regulates the dump's nonrelational goings-on. Survival of the fittest functions as the secular ideology of the fall, edging out the biological sense of fitness (to wit, the mutual adjustment of the organism and its environment[8]) and promoting the social sense of a mad race to the top of the pile at any cost, assuming that the top could be told apart from the bottom in the world-historical storm of debris.

In the theological narrative of the fall, sins individuate the sinner, whose confession of her forbidden thoughts, desires, and acts is also an invention of sort, the calling forth into existence of her subjectivity. The background for such individuation is the uniformly distributed and anonymous original sin, marking every human before birth by means of a *maculate*, stained conception. Original sin is the engine of the Judeo-Christian theological machine, with a gigantic chute spitting sinners out into the world (the dump). Earlier yet than one's origin, prior to decision-making, memory, control, consciousness, and unconscious acts by omission, to sin is to be, to have been born. Singular sins pale in comparison with the generic brand burnt on human flesh, the brand that *is* human flesh. In no way can one confess it and expiate *being itself* outside the Christian story of the immaculate conception and universal debt-forgiveness, the birth and death of god for humanity.

After the second death of god conventionally styled as *secularization*, the global dump encourages the expiation of being through a transformation of mass into energy—a pure, flammable, explosive potentiality. Existence is tolerated, so long as nothing stays still, beings are not detained within their proper limits, and being itself speeds along toward nothing. Secular ontological expiation is now contingent upon the intensification of the massive fall, which needed expiating to begin with, the fall of beings en masse. Their expiation is, in actuality, their extirpation. We have not yet swerved away from Neoplatonism.

je suis biomasse

On January 7, 2015, gunmen shot and killed twelve people in the Paris headquarters of the satirical weekly *Charlie Hebdo*. Immediately after the attack, the slogan-cum-hashtag *Je suis Charlie* ("I am Charlie") popped up on Twitter, a brainchild of Joachim Roncin, artistic director and journalist for *Stylist* magazine.[1] A hashtag and a banner in demonstrations in support of freedom of speech and of the press, the statement promptly gained currency around the world.

At first glance, the dump is anathema to everything *Je suis Charlie* stands for. The proclamation is an unconditional expression of empathy with the victims, with the institution, and the idea they represented. Apparently complete, the identification with "Charlie" exhausts the identifying *I*. Nonetheless, the digital dump imposes its own set of rules, and today "I am Charlie," while tomorrow, having erased yesterday's zeal from my mind, I am someone else altogether, with the same intensity of conviction. *Je suis Charlie* is a particular act of identification that dismantles its particularity and unravels the work of identifying with the other, even as it sneers at the prospects of concocting a universal identification. It invites an abstraction, a formal and formulaic identity: *Je suis X*. A generic variable *X* determines who I am at the moment in a wholly indeterminate way, ever ready to change (without becoming) into another thing. In a dump, limitless subjective mutability dovetails with the objective immutability of industrial by-products and metaphysical or theological ideals. Transcribing *Je suis Charlie* into anything is a no-brainer: it has no form adequate to it, which means that *I* have no form. *Je suis X*, therefore, *Je suis Ahmet, Nigeria, Nisman, Boris Nemtsov, Paris, Daphne*. . . . In sum, *I am a dump* (of identifications).

The French *Je suis* is, for all that, a highly unstable locution. As Derrida notes,[2] it can mean "I am" or "I follow," depending on whether we spot in it the present tense of *être*, to be, or of *suivre*, to follow, in first-person singular. The assertion "I am X" itself fractures the unity of the subject and the predicate it pools together: their articulation requires time (at minimum, the time needed to say or write this), cleaves the "I-X" unit by squeezing "am" into it, and testifies to the crumbling of the conditions under which it goes without saying who or what I am. As soon as we settle on the verb *suivre*, a spatiotemporal fracture between me and that which, or the one whom, I follow widens. This development is ominous at a time when following and unfollowing someone is done with a simple click of a button on Twitter, the original medium of *Je suis Charlie*. Why? Because "[i]t is necessary to know *how* to follow others, so as to emancipate oneself from this somewhat subservient relation. And yet, digital following precludes the active component . . ., as we are drawn along by whatever is 'trending' at the moment. The more we practically follow others, the less we know how to follow, or what following means."[3] Whoever or whatever it is we follow in this obsequious manner leads us directly to a dump.

A sincere identification at the current historical junction is not one that insists on the uniqueness of an ethical bond to the other, ergo on the impossibility of truly identifying with her alterity. Nor does it hinge on a fluctuating, trending or trendy Cause, followed with a semblance of ardent passion and profound empathy only to be abandoned the morning after. A brutally honest identification would be consciously, explicitly, purposefully deindividuating, mindful of its dumpiness: *of* the dump, not *for* the dump. Hints of transcendence glisten in the immanence of the genitive "of." Cultivating an identification *of* the dump is embodying and enthinking the dump from within, so as to break its spell. Isn't discerning the dump, and, thereby, discerning what prohibits discernment, the only way to peer out it?

This context proves fecund for a reflection on another, more recent (and rawer, too), hashtag-turn-political-campaign. I have #metoo in mind here. The denunciation of and punishment for acts of sexual harassment, assault, and rape under the banner of the #metoo campaign are long overdue. Yet, the crucial question is not so much *what* is going on as *in what ways* this watershed event that augurs a revolution in cultural perceptions is taking place.

The identification that says *Me too!* is plainly not singular, its generalization peculiar, to say the least. If "too" combines multiple victims of rape, assault, and harassment, it does so by accretion and agglomeration, compiling, piling up heterogeneous experiences in an undifferentiated whole. As various commentators have stated, the degrees of severity in the gamut of sexual misconduct tend to become blurred when illuminated by the #metoo movement. Unexamined, the problematic of following we have glimpsed hounds at every turn the movement's formation and growth.

The political subjectivity stemming from #metoo is as blurry as the demarcation lines between unwanted sexual advances and forced sex: passive identification with an amorphous Other, who is also (like "me") a Victim, repeats the passivity of having been victimized by an equally amorphous Other, who is a Perpetrator. The apparatus of identification that works by adding me (too) to the heap of victims does not address the need to work through victimization, rather than let oneself be engulfed by the flames of reactive affects (rage, resentment, etc.) it ignites. After all, the political quandary *proper* has to do with the steps one needs to take in order to transition from *me* to *I* and to *we*.

Come think of it, *me* is a weird personal pronoun. On the one hand, it corners the subject into a purely passive object—in this case of sexual harassment and kindred types of behavior, as well as an object for dumping onto a pile where the others are, *too*. On the other hand, it is the pinnacle of a narcissistic attitude and, above all, of melancholy narcissism, also marking the navel-wound gazing of the Anthropocene. A narcissist is paradoxically incapable of saying *I*. For her or him, it is always *me, me, me*: the world revolves around me and, to retain its relevance, has to, in one way or another, act upon me, who does not stand as an *I* in relation to it. Any campaign feeding off and fostering such a narcissistic stance will end up in a political cul-de-sac.

Succinctly put, #metoo is an instance of a deindividuating identification, albeit one that is not consciously embraced. It replicates the main traits of the dump: the sheer massiveness of accusations; the rapid drop (also en masse) of the accused, whose fall bursting at the seams with religious connotations is supposed to elevate the victims to moral heights; the piling of bodies, both of the violated and the perpetrators; disregard for the time frames *before* and *after* the start of the campaign. If a social and political movement—whether it is #metoo or the majority of new populisms—is *of* and *for* the dump, then the doubling of its origin in the destination, the folding of the origin unto itself, dramatically overshadows its revolutionary potential.[4]

Walt Whitman's *Leaves of Grass* is a good literary antidote to *Je suis Charlie* and, perhaps, to the unresolved contradictions in #metoo. The hymn gives voice to the self-consciousness of the dump, inasmuch as it puts opposite qualities on the same footing. For instance: "I am of old and young, of the foolish as much as the wise, / Regardless of others, ever regardful of others, / Maternal as well as paternal, a child as well as a man, / Stuff'd with the stuff that is coarse and stuff'd with the stuff that is fine . . . ."[5] To the formula *I am X* Whitman prefers *I am of X*, a genitive that promises transcendence. "Of the old and young, of the foolish as much as the wise" makes me simultaneously into less than each group, of which I am a part, and into more than the whole I belong to, thanks to an open-ended chain of other belongings. The preposition *of* keeps in check the totalizing power of identification. Resembling an incantation, the poem stockpiles disparate determinations *of X and of Y*, the latter habitually thought of as *not-X*, neither conserving their schisms nor forging higher syntheses out of them. I am, Whitman circuitously suggests, a dump of the most diverse qualities, plus the cognizance of their amassing.

The conjunctives *and*, *as well as*, *as much as* fashion me into a *complexio oppositorum*, an alchemical and, according to Carl Schmitt, a theologico-political dump.[6] The *complexio* compiles opposites immediately, without mediations and delays. (We should say *piles up*, rather than *compiles*, because *com*pilation is indicative of a prior mediation.) Although a lag between them necessarily remains, if only as a result of interposing a conjunctive, the unmediated collation of opposites discloses a truth that takes some time to unspool in *Je suis Charlie*. The truth of the statement is its negation: *Je ne suis pas Charlie* ("I am not—or I do not follow—*Charlie*").

An exceptionally helpful indication Whitman gives his readers is germane to matter, to the materiality of who or what I am. "Stuff'd with the stuff that is coarse and stuff'd with the stuff that is fine," I am mass ("coarse") and life ("fine"). To paraphrase: I am biomass, a kind of heaviness heaving with life thrown as a shapeless lump into death-bearing energy production. I am a pound of flesh, or a hundred fifty pounds of it. *Je suis biomasse* is a deindividuating, anonymous identification testifying to the precarious situation of living death on a global dump.

Biomass is primarily vegetal: wood, and crops (corn, sugar cane), and alcohol-based biofuels (ethanol) derived from plants. But, along with vegetation and on par with it, the word comprehends organic garbage and landfill gases (methane), animal manure, and human sewage. What these diverse components have in common is that, akin to conventional fossil fuels, they are expected to go up in flames so as to supply energy, even if, unlike coal, oil, and natural gas, they are classified as renewable resources, springing forth as though from a limitless stockpile. There will always be more plants, municipal waste, and excrement where these have come from. . . Is their affinity to the self-renewing nutritive function of life a good enough justification for massively dumping the residues of their conflagration into the atmosphere? In effect, the resulting CO_2 emissions often quantitatively surpass the consequences of burning fossil fuels. The biomass portion of renewable energy warps renewability into infinite (with regard to the source, not the consequences) dumpability. It is unrelated to wind, solar, hydro, or geothermic power that bespeaks the need and the desire to work with the elements in a synergistic relation constitutive of energy.

A perfect fit for the global dump, biomass is ageless and born aged. It shuffles and reshuffles disparate temporalities until all differences among them are eliminated: the growth and decay of its living and dead ingredients, the volatility and the relative stability of gases and wood, the time it takes to cultivate a crop and to transform it into excrements or into a pile of burned stuff. Not only can we not say with confidence *what* it is, but the question *when is it?* remains unanswerable. In this respect, Marshall McLuhan's quip upon seeing a typo in the title of his *The Medium Is the Message*, which came back to him from the publisher as *The Medium Is the Massage*, is instructive: "Leave it alone! It's great and right on target!"[7] Why? Because McLuhan grasped the promise of describing the Information Age as a "mass age" and as a "mess age," the age of the masses and of messes. Even more so, biomass (a life mixed with nonlife) has its age (the rhythmic unfurling of finite temporalities) massified and tangled into an impenetrable mess.

The ingredients of biomass demonstrate with crystal clarity that, in a dump, plant life is interchangeable with (useful, i.e., burnable) garbage and shit. To be sure, attitudes of the kind have been widespread for centuries, but never before have they been spelled out with such nonchalance. Lest we have the impression that vegetation is alone in this predicament, we should take a second, harder look at the conceptual skeleton of biomass. Biomass absorbs life into weight and, constraining it to formless matter, obliterates it in advance of biological extinction, still before actual species and biodiversity losses. At the crest of mass society, a shared, social and political phenomenon of *bios* withers away. Contrasts between modes of life, between the "sociality" of *bios* and the "natural" vitality of *zōe*, peter out, with everything eventually reduced to the lowest common denominator, the mechanics of force, gravity, the fall of that which has and is mass. Distinctions among lifeforms, within forms of life and between the forms *life* and *nonlife*, become irrelevant: all are presumed lifeless. Necropolitics is not at the antipodes of mass biopolitics; it emanates from biomassification.

Vegetation is at the forefront of biomass. Since Aristotle, its vitality has been linked to a generic mode of living shared by plants, animals, and humans, malleable and exuding potentiality in excess of organismic logic. Plant life, reconstructed from the perspective of formal-logical and abstract reason, was little more than a dump. (A psychoanalyst would say that the rational reconstruction of vegetation in a negative light is indicative of reaction formation on the part of the rational human subjects themselves.) To combine *bios* and *mass* is to read vegetal plasticity between the lines of these ancient protocols.

But biomass is, more pointedly, a trashcan term: on top of the actual plants, nonvegetal organic waste is hauled into it thanks to the association of muck and the like with the vegetal nutritive faculty, *tō threptikon*, and, through it, with a living-breathing-eating-shitting body. While that body may be of a plant, an animal, or a human, a deindividuating identification with and as biomass puts us in proximity to plants that, here too, are at the cutting edge of the latest historical permutations of being.

The fate prepared for all lifeforms is to be biomass, muddled organic junk without the time and being of their own. It is also to go up in smoke on the pyre of energy production. The fumes rising from the planetary blaze are a travesty of matter's negation, suspension, elevation, relief, overcoming, preservation, sublation, sublimation, spiritualization (and any other senses of Hegel's *Aufhebung* I may have left out). A worldwide conflagration turns the sky into a dump for carbon emissions. The yellowish-brown hues of nitrogen oxides and the grey pallor sulfur dioxide casts over the areas it affects hand the air over to sight as smog. Ethanol combustion yields carbon monoxide, carbon dioxide, and aldehyde, which promotes photochemical reactions. Becoming visible, the aerodump reduces visibility and eats into our lungs. We grow and burn entire crops of sugar cane and corn alive in a deranged quest for energy. Without regard for the living, the dead, and the razor-thin line between them, we burn at the stake the biomass that they are in nothing less than the Vegetal Inquisition. The Middle Ages saw human heretics incinerated on bales of wood and dry straw on the pretext that fire would cleanse their sinful but immortal souls as it consumed their physical bodies. Nowadays, the plants themselves are burned in a secularized drive to atone for their biomassified formlessness by acceding to the only shape that befits matter as mass. That shape is energy. Modern metaphysics converts matter into mass before physics reconverts it into energy. Things are straightforward enough with respect to potential energy stored in a physical system before its release and transformation into other forms of energy. What about Einstein's equation $E = mc^2$, though? Can we revise it for biomass? Does *to be* equal being dumped at the speed of light, squared?

Plants are not alone in the flames. No sooner does biomass materialize out of the erstwhile metaphysical perspectives on matter than partitions between us, who burn something or someone, and them, who are burned, melt away or disintegrate into ashes. We are the incinerated incinerators, the dumped dumpers, inflamed by the cinders of a desire that remains opaque to us, breathing in the fumes of deadly energy production, and throwing ourselves on the pyre together with our world. In turn, I am biomass: the ephemeral and flammable heaviness of matter still heaving with life at the crossroads of a living mass and massified living. *Incipit* a long passage from *me* (too) to an *I* and a *we*!

In his nascent theory of the state, Nietzsche hypothesized that the masses could be shaped by the political "artists of existence" who "continued to work until finally such a raw material [*ein solcher Rohstoff*] of people and half-animals was not only thoroughly kneaded and pliable but also *formed*."[8] The creation of social and political forms (*Formen-schaffen*) hinges on their being violently impressed (*Formen-aufdrüken*) onto the raw matter of the population.[9] The ultramodern gist of Nietzsche's remarks is that political form does not grow organically from the materials it organizes. It is not absent altogether, however, seeing that it is utterly indispensable for the task of articulating pacified, docile, pliable human or half-human masses with determinate shapes of collective existence in the unity of the state.

Toward the end of the twentieth century, mass production, mass societies, mass media, and mass communication (the new four horsemen of the apocalypse?) jubilantly realized a disconnect between political matter and form.[10] Mass society culminated in mass dissociation. To put the masses to work, to draw their energy, suffice it to dump them down to earth, capitalizing on their heaviness, or to incinerate them and dump them up into the sky. A massive throw thwarts the political humanization of the masses, which Nietzsche explores via the formative mold imposed on human half-animality, as it blends them with the rest of the biomass fueling the furnaces of capital's economy. Human resources are raw materials, expendable and easily available, formed exclusively outside themselves in the material or immaterial commodity they help manufacture. Though burned or suffering from burnout, they are renewable, drawn from a never-ending stockpile together with corn crops, garbage, and excrement in the most recent installment of surplus value. And this is not to mention the vast, devastated regions of the planet where the starving human masses excluded from the circuits of mass production, society, media, and communication are not disposable but disposed of, denied all but a marginal connection to *bios* and about to lose *zōe*, the life, for which Giorgio Agamben reserves the adjective *bare* in those circumstances when one is stripped to one's animal (and vegetal) metabolic or physiological vitality.

The biomassification of humanity advances on two fronts: 1) *bios* comes unglued from a shared language and action that, in Hannah Arendt's view, are the preconditions for a new beginning in our second, political birth; and 2) the *masses* of production, society, media, and communication alienate, isolate, and disarticulate those dropped into and amassed in them. On both fronts, the effects of human biomassification are comparable: heaps of bodies and thoughts; materials and narrative fragments that are no more than caricaturized types and isolated traits; objectified labor and lifestyles, those poor substitutes for a form of life. All this accrues fragmented, shredded, dismembered, disjointed, atomized, analyzed, but not decomposed.

The two-pronged process is similar to how fire breaks things down, even if the legacy of a biomassed dump is the earth not strewn with fertilizing ashes but covered in toxic garbage. Novalis drew a telling parallel between mind-body and fire-matter relations: "A problem is a solid synthetic mass, which is broken up by means of the penetrating power of the mind. Thus, conversely, fire is nature's mental power and each *body* is a *problem*."[11] A part of biomass, thought and its workings are solid synthetic units, dissected and desiccated especially at the hands of a philosophy known as analytic. Dried up, processed for and by the desert, they are ready for setting alight. The mind, too, is subject to alienation, isolation, and disarticulation, its rests cast into the global dump. The algorithms that condense cognitive functions and decisions into a formula are the sterile ashes of thought's incinerated biomass. After a while, you will find no trace of synthetic solidity there, nor of imagination, space- and time-consciousness, experience. Only a muddle of thoroughly analyzed, carbonized cognitive scraps.

(I note, in a telegraphic manner certainly worth unpacking on another occasion, that the algorithmic regulation of everyday life operates in the space of a basic, if unnoticed, contradiction. In our choices of books, vacations spots, or romantic partners, algorithms know us from within, much like Augustine's god, who resides also in the human soul. They are thus finely and infinitely differentiated. But they indifferently work themselves out on the undifferentiated grounds of pure quantity, spurning all other categories. The algorithmic manipulations of information craft a world corresponding to each of us *and* pertinent to a dump, wherein the world is impossible.)

The flaming transmutation of matter into biomass fails to provide any mediations between being and nothing; it fails to become what becoming is in the Hegelian system. According to *The Science of Logic*, becoming commences with the negation of the proposition "being and nothing are the same."[12] In *Encyclopedia Logic*, "the principle of *becoming*" is "that being is the transitioning into nothing and nothing the transitioning into being [*das Sein das Übergehen in Nichts, und das Nichts das Übergehen ins Seins ist*]."[13] In a dump comprising undecomposable fragments of beings, this two-way transition comes to a grinding halt and the proposition "being and nothing are the same" is valid again. But, there where dumped biomass is biodegradable, the transition continues in a terribly lopsided fashion, moving from being to nothing, not from nothing to being. Already underway, the sixth mass *extinction*[14] is a massive repercussion of the mass *extinguishing* of life at the end of its mass incineration and the erratic dumping of the earth into the sky, above all, by burning the long-dead, the fossils. The command from Revelation 1:19—"Write . . . that which will become [*ginesthai*] after this"—may be unfulfillable, seeing that, already as of this writing, becoming is itself becoming a thing of the past.

"I am biomass" is a speech act that identifies with a vanishing life, with life's vanishing into dumped massiveness. The sentiment behind these words is ineluctable: one must have been dumped by being to resist the dump. But what if *Je suis biomasse* actually means to say "I follow biomass"? Do I come after it, becoming a survivor and hijacking whatever remains of becoming? Do I merely accompany its trials and tribulations in the newsfeed? Am I fleetingly interested in its fate, which I take to be unrelated to mine, ever ready to shift my attention to something else the next moment? *Is* there anything or anyone exempt from the dynamics of biomassification?

In the wake of these questions, we ought to steady, albeit provisionally, the semantically unstable expression *Je suis biomasse*. A decision for *to be*, *être*, rather than *to follow*, *suivre*, is called for. It is necessary, of course, to understand identification with biomass. But it is more vital still to sense this identification in one's bones, those repositories of dioxins, lead, and, in radioactive fallout areas, of strontium. To be biomass, regardless of its aversion to being. What does it feel like for *I* and *am* to be in proximity to *biomass*? From the point of view of the identifying subject, it is akin to the experience of falling into a black hole in an incessant approach to the event horizon that appears to be ahead of the one who has already crossed the threshold. The affirmation says: I am decimated being and stymied becoming, yet not exactly nothing. Dumped, I resist the dump with the surreal power of not-nothing, the power allied on the olfactory level to the stench of rotting.

antilogos

The dump is a site for the amassing of whomever and whatever falls into it. Yet, it is not a place. The dump does not take place and does not assign to the fallen their proper places. Despite their physical proximity, its sundry parts do not pull together and do not draw apart. A toxic cocktail of disengagement and homogenization nullifies relations: each is too far from and too near to the other. The continuum of sameness and difference closes into a circle at its extreme ends (pure sameness, pure difference) unyoked from relationality. This circle is the circumference of the dump's mouth, which is also its anus, infinitely multiplied and spread throughout its extension. The dump proscribes gathering whatever passes into it. Its pieces are unlike *Lego* blocks one can use to assemble various structures. The heap that it is has no *logos*: it is without articulations in the double entendre of a vocal expression and spatial jointure at the hinges. Nothing can come out of it, not even compost.

The dump is unhinged and unhinging. It stops every pursuit of an assembly, every effort of regrouping, regathering, and community-building in its tracks. In enigmatic fragment 124, Heraclitus, the pre-Socratic mouthpiece for *logos*, collocates a chaotic heap with a beautiful world-order and a shimmering decoration that is the ancient *kosmos*. Not one to separate opposites, no matter how polarized, he writes: "Just as a heap of refuse piled up without purpose, so [is] the most beautiful world-order [*all hōsper sarma eikē kechumenon ho kallistos kosmos*]" (D-K 22, B124).

What does the fragment say?

It is an analogy, its parts—a dump (*sarma*) and *kosmos*—standing in a definite proportional relation toward one another. This is not a minor detail. Translators have tended to skip over the words "just as," *all hōsper*, and inverted the fragment's word order, implying that what it is really about is *kosmos* defined as a heap of refuse. Their puzzlement is understandable. Heraclitus comes up with an impossible analogy, a proportion between disproportionate things, or, to be exact, a proportion between perfect proportionality and the want of proportion, the ana-logy of *logos* and *alogos*. After all, he would not have been the thinker that he was had he not articulated these poles with "just as."

An extra layer of complexity surfaces with the realization that Heraclitus constructs his analogical fragment in the form of a twofold tautology. To translate again: "Just as a dumping dump, so [is] a cosmeticizing cosmos." The tautological arrangement did not escape readers such as John McDiarmid, who in the 1940s took the inclusion of the meaning "the most beautiful" in *kosmos* as a cue for omitting this word on the pretext of its redundancy and rewriting *sarma* as *sarx*, "flesh." His proposed translation was: "The fairest [of men] is flesh scattered at random."[1] As a consequence of McDiarmid's abrupt interpretative move, the sense of the fragment was domesticated and perfectly conformed to the ontological and theological frameworks subsequent to Heraclitus, where flesh was, indeed, a dump for sentient matter.

Discomfort with tautologies is a modern attitude. It is difficult to ascertain what repetitions in "a dumping dump" and "a cosmeticizing cosmos" have to contribute to our appreciation of the fragment. As a matter of fact, they bear an invaluable gift for understanding. Invoking both a *how* and a *what*, Heraclitean tautologies ferret out the energy of being, which is an entity as well as an activity, a noun and a verb.[2] Could the working work of energy be the true subject of the analogy? *Sarma* is the substantive outcome of the act of dumping, of piling up (*kechumenon*) what has been thrown aimlessly, without consideration, at random, or in vain (*eikē*). The dumped is essentially hidden from view, the heap obscuring its contents. A miscellany of non- or antiphenomena, *sarma* is not itself given; it names this very flight from givenness. The beauty of *kosmos* is, on the contrary, the substantive effect of shining cosmic activity, "kindling in measure and extinguishing in measure" (D-K 22, B30). Phenomenal through and through, it delivers itself to sight, with measured, rhythmic, regular intensities of fire dictating the order that at the outset strikes the eye in the form of beauty, if not in form *as* beauty.

The two tautologies thus stand in an antithetical relation. But they are also analogized: the withdrawn, disorganized pile is "just as" the available, shining order. Why? First, their proportion is not preordained. *Even* a randomly heaped pile *can* in due time metamorphose into the most beautiful world-order. Just think of manure, in which a charming flower may blossom! Second, an antiphenomenon is unsuccessful in sidestepping the exigencies of givenness and self-givenness. However vague the outlines of the dump and however inimical to being, they are apparent, which is why, incidentally, a depiction of its amorphousness is not a contradiction in terms. Third, the pile may contain a latent principle of arrangement undisclosed to physical or mental sight. A rift will then open between the two senses of *kosmos*, namely order and shining beauty, the former present without the latter. Fourth, the glow of world-order is not constant; it "kindles in measure and extinguishes in measure," so that the phase of extinguishing, descent into a heap, unworlding is not the direct negation but an intimate possibility of the Heraclitean world. Fifth, and relatedly, the dual character of energy boiling in the tautologies is antithetical: unrest and rest, an activity and its outcome wherein the act's agitation is quelled. Far from a constant and inexhaustible blaze, energy evinces a measured lighting up and dimming down of the world of ancient physics.

Fast forward to the global dump of the twenty-first century. Here, the vicissitudes of kindling and extinguishing are destitute of measure. It is not that an objective set of limits and the virtue of moderation they prop up have been lost, as a conservative will, no doubt, argue. The measure is not quantitative, let alone sneaked in from outside that which it measures; rather, it is internal to Heraclitus's fire and to phenomenality. A cosmic barometer, it gauges the fluctuations in pressure that sustain the increase and decrease of elemental heat and luminosity. These fluctuations have become extreme to the extent that the two "powers of fire" have been dissociated from one another, converted into heat without light and light without heat, pure passion and pure reason.[3] In the history of being, endless extinguishing presents itself as our fate. Rekindling becomes ever more difficult and, when it happens, resembles something like spontaneous combustion.

An interruption of successive kindling and extinguishing means, beyond a glitch in the rhythmic alternations of apparent ordering and nonapparent disordering, that the dump has been set loose from its restricted role as the eerie underbelly of a beautiful world-order. The unworlding of the world is no longer *of* it in the sense of a moment proper to a world it helps restructure and renew. A growing heap of refuse smothers, cleaves, and diminishes the liveable regions. At their expense, *sarma* and the extinguishing that defies rekindling wax immeasurable and unfathomable. That is the key motive for switching parts of the fragment around and erasing its opening words in numerous translations. Heraclitus sought the analogue of a shimmering world-order as low as in a garbage heap; we manage to twist and grate the most beautiful *kosmos* into a dump.

If *kechumenon* is the antiphenomenon, then *sarma* is antilogos, and together they impede phenomenology. More, their energy is anti-energy. I am not denying that they house an empowering positivity, the very power of power, the potency of potentiality. Piling things up foreshadows the appearing of phenomena, putting visibility at loggerheads with the invisible; a disorganized pile is the condition starting from which *logos* initiates its activities of gathering, articulation, assembly. *Sarma* is the chaos one must have in oneself so as "to be able to give birth to a dancing star [*einen tanzenden Stern*],"[4] a celestial body that reflects the ancient shimmer of *kosmos* and, letting it "dance," dynamizes world-order. It is not the chaotic heap per se that is worrisome but its declaration of independence from the nets of articulation torn apart by falling massiveness. With this declaration, the dump holds back the building blocks for a meaningful assemblage of the world, the *Lego* pieces for a provisional (always only provisional!) order. Hypostatized, *sarma* is the other side of the dialectical moon, a nonrelation of relation and nonrelation.

Heidegger hears in the fragment by Heraclitus a strange, disconcerting saying, *eine befremdliche Wort*. In *Introduction to Metaphysics* he continues: "*Sarma* is the counterconcept [*Gegenbegriff*] of *logos*, what is merely cast down as opposed to what stands in itself, the heap [*Gemenge*] as opposed to collectedness, unbeing as opposed to being. [. . .] Gathering is never just driving together and piling [*Zusammentreiben und Anhäufen*]. It maintains in a belonging-together that which contends and strives in confrontation. It does not allow it to decay into mere dispersion [*Zerstreuung*] and what is simply cast down."[5]

Sarma falls *down* into a heap; *logos* stands *up* straight. The former disperses; the latter maintains a belonging-together. *Logos* erects the thing *sarma* causes to fall, even if in its monumental massiveness the pile of that which has fallen imitates, pillar-like, *logos*'s erection. They are the topsy-turvy renderings of one another in space and in time. All the same, Heidegger's chief preoccupation is their opposition at the level of the concept. A counterconcept, *Gegenbegriff*, means that one concept is pitted against another *and* that one of the parties in this deadlock (namely, *sarma*) works against the logic of conceptuality, which is the logic of *logos* as well. Conceptual mediations are extraneous to the counterconcept that im-mediately heaps the opposites it paradoxically gathers into a pile and scatters. (Still, the concept anticipates the creation of a dump, as it decontextualizes, abstracts, and indiscriminately lumps together particular instances of a *this* under its umbrella.) To speak of oppositions is already to stand outside *sarma*'s unworld, on the solid terrain of *logos* and the concept. It is to stand up against all odds of falling down. This position, this upright posture preserves the Heraclitean analogy and imputes a relation, if a negative one, to the extremes. The oppositional stance leads us to a conjecture that what the counterconcept opposes (being, collectedness, standing-in-itself . . .) is the default state, from which the negation deviates. Anyone who assumes such a position in thought or in deed subscribes to the unity of being and unbeing, *Sein und Unsein*, of a world-order kindling in measure and extinguishing in measure, the congruence of its positive and negative moments validated in time.

Nonetheless, Heidegger receives the disconcerting strangeness of Heraclitus's fragment by capturing *sarma* in the word "unbeing," *Unsein*, instead of the more usual "nonbeing," *Nichtsein*. Unbeing is not a negation of being, or, at least, it is not just that. Having abjured relationality, unbeing is the actively nihilistic undoing of being that is not locked in mortal combat with being and has nothing to do with what it undoes. The disengagement of "nothing to do with" accomplishes the work of radical undoing. Strictly speaking, one starts saying nonsense as soon as one avers, as we did in Heidegger's company, "*Sarma* is . . .": "*Sarma* is the substantive outcome . . ."; "*Sarma* is the counterconcept [*Sarma ist der Gegenbegriff*]," and so on. The word *unbeing* corrects the oversight by insinuating that *sarma* neither is nor is not. In its substantive form, the dump desubstantivizes, deformalizes, and deforms itself, exhibiting no connections to the being *that* is and to *what* it is—in other words, to Aristotle's first or second *ousia*. The *sarma/kosmos* analogy in the original fragment disintegrates. On retreat from being and becoming, the dump fosters pure virtuality. The surreal power of my biomassified not-nothing must contend with the virtual unreality of the heap.

Unbeing cannot but ring alarm bells during our drawn-out party celebrating the implosion of grand narratives. It seems that the micrological storying of the world, splintered into a plethora of equally valid perspectives and accounts, has prevailed over hegemonic World History. But another intriguing speculative prospect is that, thanks to its virtuality, the dump has been imperceptibly set on the empty pedestal devoted to a master narrative, without filling the emptiness left behind by the deposed master. Where unbeing supplants being and an all-embracing reality is an unreality, the absence of a single entity, determining factor, or concept pulling the strings behind the scenes is the absence proper to the dump. The decadence of grand narratives might be the dump's *via negativa*, an oblique piece of evidence for its alliance with our apophatic theology in disguise, hostile to *logos* and to being. Henceforth, the talk-shows of the disappearing world are screened and resound in a place that can be hardly shown or talked about.

The global dump is as close as it gets to designating the ontology of contemporaneity. The puzzling bit is that it does not amount to an ontology, perhaps not even to Derrida's *hauntology*, a spectral play of presences and absences.[6] *Sarma* dilutes *logos* in logistics, the proceduralism of dumping, into which the law sporadically intervenes: No illegal dumping! (Plenty of the legally sanctioned kind is happening, though.) Once discerning voices, speeches, and articulations fall silent, the air fills with the ear-splitting roar of existential dump trucks and the loads they thunderously drop. Pine as we may for silence, the aesthetic soil wherein *logos* may again germinate, this is a pipedream at best and irresponsible escapism at worst. The only way to wrap oneself in "peace and quiet" is to plug one's ears, keeping out the noise of the massive fall that, in turn, drowns the screams of the dumped. Wouldn't one lie primarily to oneself then? Not hearing oneself speak and not hearing oneself hear, one would not as much as hear oneself fall.

toward an intellectual history of heaps, piles, and other jumbled things

Heraclitus and Walter Benjamin are the bookends of the philosophical proclivity for thinking the dump—for thinking, therefore, on the peripheries of thought. A confused pile, into which everything and everyone is thrown, this antiphenomenon ensconced in antilogos both repels and attracts philosophers from Aristotle to Kant, Plato to Hegel. The dump is sublime. By their vocation, philosophers defy it: some, because it frustrates their desire for order; others, because, true to their *love* of wisdom, they detest the apathetic disengagement the dump foments. But even the staunchest and most ardent proponents of order agree on two things: (1) an unruly pile of stuff is how the world is first given to us (as well as how we are given to ourselves in infancy), and (2) chaos, involving potentiality in general, contains among other things the potentialities of order.

Enamored of unrealizable possibilities, detained indefinitely at square one where the actual always falls short of the ideal, receiving unwieldy information beyond our receptive capacity, we are barred from moving past the world's first givenness that borders on the nongiven. We are hedged in a "great blooming, buzzing confusion" of a baby "assailed by eyes, ears, nose, skin, and entrails at once," as William James puts it.[1] The world is dumped on us while we are dumped into it, which means that we face the task of painstakingly, bit by bit, individuating the given and, in so doing, individuating ourselves. On a global dump, this is literally a nonstarter: confusion persists inside and outside, between the inside and the outside not deconstructed (and, in the course of being deconstructed reconnected—without any solid guarantees of success—to the movement of recurrence, reanimation, and finite preservation) but devastated in humanity's "mere oblivion" and "second childishness."

Whereas James's blooming, buzzing confusion has become an intellectual buzzword, the context of the phrase is rarely mentioned. Addressing the unity of objects, James presents incipient experience as a dump of sensory impressions: *"any number of impressions, from any number of sensory sources, falling simultaneously on a mind which has not yet experienced them separately, will fuse into a single undivided object for that mind."*[2] The indifference of (any) sensory inputs and of the number of impressions on the sensorium, the fall, simultaneity, massiveness—every quality of the dump is prominent in this description. A "single undivided object" is a corollary to experience that has not yet been individuated, the upshot of a mind bombarded with indiscriminately fused impressions. The confusion that the undifferentiated and indifferent fall of impressions provokes is magnified by our incapacity to receive the given assailing us on all sides, all the time. All the sides, on the inside and on the outside, add up to no side whatsoever; *at all times*, irrespective of periods, phases, or sequences, means *at no time*. In so stamping out the places and times of experience, the absolute exerts its unbearable pressure on a living body and finite cognitive-perceptual faculties.

For James, not only the sense of an object but also mental construction of space obeys the same rule: "and to the very end of life, our location of all things in one space is due to the fact that the original extents or bignesses of all the sensations which came to our notice at once, coalesced together into one and the same space."[3] The heaping of sensations, and in particular of their magnitudes inaccessible to a purely sensuous judgment, draws a veil over the multitude of objects *in one object* and the plurality of spaces *in one space*. Alongside the dump of the mind, external space is likewise a dump, so long as the simultaneity of "at once" (that, for Kant, constitutes it) overrides temporal punctuations, be they as trivial as the jaggedness of a succession accentuating the minimal discreteness of some conjoined moments. Discernment is temporalization, made possible by ruptures in the relentlessness of space, itself a side effect of the original blur of sensations. The dump, for its part, is atemporal, untimely and untimed, despite its vast historical range, jutting peaks, and low-lying valleys. It is aporetic: there are no passages in it, no way through it, no breathable pores, no becoming, no transitioning from the first— immemorial and undiscerning—to the second givenness.

An intellectual history of heaps, piles, and other jumbled things begins (obviously, well before James's *Principles*) with the acknowledgement that the world is smashed to smithereens when the psyche is disarticulated. The world's undoing depends on what comes to pass or doesn't come to pass in psychic life, which, for its part, lives off (and also dies off) the world. As he comments on the Tower of Babel, Philo observes that at present, in the now, *nun*, "we have all of the soul in inextricable confusion [*sugkechumena ta panta tēs psuchēs*], so that no clear image [*eidos*] of any particular kind is discernible" (*Conf.* 84). In his melding of Athens and Jerusalem, of Greek philosophy and the Hebrew Bible, the Heraclitean pile superimposed onto the Tower of Babel morphs into a mental dump. The time of the dump is now, every now in which the soul unconsciously lives.

Philo's *kechumenon* is a piling-with (*sugkechumena*) of languages, psyches, materials (water, the earth, and fire in bricks of clay that go into the construction of the tower), projects, and so forth (*Conf.* 83). The Latin *con-fusio* is the exact rendering of that with-fusion. *De confusione linguarum* construes Babel in terms of a dump, where the medley of languages is but a tip of the iceberg. In its unending "now," kinds, types, species are indiscernible. Without a sharply focused image of languages and physical materials, psyches and projects, we have no feel for the being of language, the being of materiality, of the psyche, of a project. The gigantic erection of the tower is, in itself, its massive fall still before its actual destruction. *Psuchē* occupies a special place in this mess, for it epitomizes a double image deficit: we have no image *of* it and *in* it. The psyche is the dump par excellence, an abysmal dump within a dump, its interiority the identical twin of exteriority.

Husserl sifts through the psychic dump and fights against its obscurities with the entire arsenal of phenomena and *logos* at his disposal. The individuation of consciousness by its object, its singularizing intentionality as a consciousness of . . ., is a dump-proof strategy that emphasizes primary cognitive differentiation and

Figure 4 Drowning in, Anaïs Tondeur, 2018–20, Pigment print on Murakumo paper, 42 × 63 cm

non-indifference. For a phenomenologist, consciousness is always a *this*, whittled by *that* which is intended; it is, then, a reincarnation of Aristotle's first *ousia*, while the intended target is a variation on the second *ousia*. Husserl is also astute enough to account for a massive and nebulous halo overhanging and surrounding the zone of attention, nourished in its singularizing dynamism by its vague milieu. In the sphere of judgment, the dump of nonexplicit meanings situated just outside the margins of a mindful comportment may help structure the temporal flow of understanding. One can always revert to its obscure horizon so as to bring new significations to light. The result is a confusion of the distinct and the confused: "Naturally, confusion [*Verworrenheit*] and distinctness of judging can be intermingled [*miteinander mischen*]; as they are if, when we are reading, we actually and properly perform a few judgment-steps and sequences, and then let ourselves be carried along for a while by the mere indications belonging to word-formations."[4]

Potential order inheres in the mental dump, assuming that "cognitional striving *tends from 'confusion' toward distinction* [von der 'Verworrenheit' zur Deutlichkeit]."[5] A hundred years after Husserl, the cognitive drive (*Erkenntnistrieb*) is blocked, and the information dump submerges confused thought processes in a flood of "mere indications belonging to word-formations." The blockage affects both intentionality (consciousness of . . ., striving toward that of which it is conscious) and the pulsion of the unconscious, the psychoanalytic drive, *Trieb*. The drive is stuck; fixation ensues.

Paradoxically, the stuckness of the cognitive drive culminates in a fixation on a permanent flux and confusion, uninvolved with the "distinctness of judging." Our attention is oddly unfixed, transfixed by turbulence and instability. We are swamped with permanent dazedness and distraction, a clamorous celebration of the chaos and uncertainty principles, indeterminacy and nonactualizable potentiality. We live directly in the halo of experience, which does not lend itself to being experienced, and miss out on what is surrounded by it. Mere indications are the horizon of mental life and fuzzy silhouettes appearing against that horizon; nothing else gets through.

A fixation on the permanently unfixed also elicits the opposite reaction, the pull of nostalgia for the presumed certainties of the past. The reaction, in turn, is a far cry from the "distinctness of judging"; it craves prefabricated clear delineations authoritatively handed over from on-high—the certainties of God, nation, and family, to put it in a lapidary way. The global resurgence of populism is greatly indebted to a collective wish to emerge out of the halo of experience. But all that it recovers is an empty husk, already fossilized and unviable, no longer supple, if it ever has been.

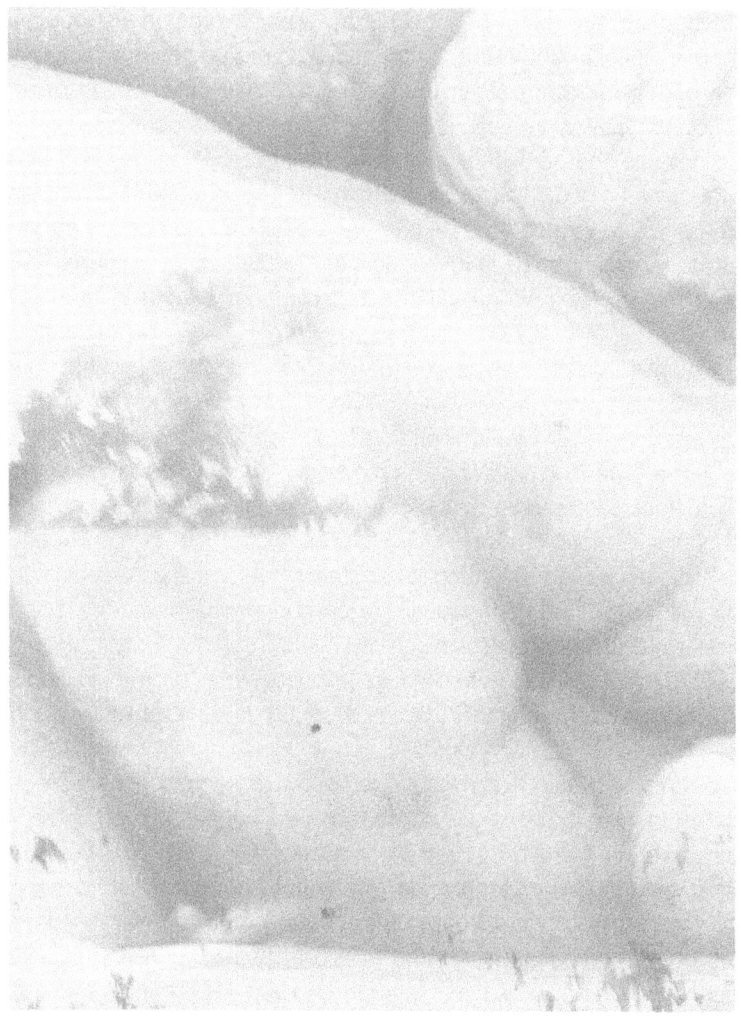

Figure 5 Drowning in, Anaïs Tondeur, 2018–20, Pigment print on Murakumo paper, 42 × 63 cm

For Plato, the mind's dump is the confusion of the senses, the fusion of their "data," which is not quite synaesthesia. Differences in magnitude are indistinct (*achōrista*) from the perspective of vision acting on its own: "Sight too saw the great or the small, we say, not separated but confounded [*sugkechumenon ti*]. . . . [T]he intelligence is compelled to contemplate the great and small, not thus confounded [*sugkechumena*] but as distinct, against sensation" (*Rep.*, 524c). The senses dilute categorial differences. Irrespective of the sharpness with which vision sees great and small things, for it, the distinction between smallness and greatness is blurry (or should we say *enigmatic*?). The vision of the categories cannot rely entirely on physical sight; it must be informed by eidetic seeing.

The category of quantity—and, with it, the entire cognitive apparatus of understanding—is not so foreign to sensation in Plato. But, although sight *sees* the great and the small in the things endowed with these characteristics, it treats the one as the other, shrinking greatness to smallness and inflating smallness to greatness. The ocular sense, in line with the rest of the senses, is a dump for the extremes it levels down. Noetic interventions work against sensation and with it; against-with, they refine the differences that, registered by the senses alone, commingle to the point of sameness. Left to its own devices, however, the confusion of seen magnitudes swells or shrinks beyond measure. In this *beyond*, one cannot tell contraction from increase. The massive is the minuscule, two count as one, and "each is one and both two" (*Rep.*, 524b) in the dump of confused impressions.

So potent is the nihilism of the dump that it invades our sensorium. Kant agrees with Plato that sensory impressions without intellection descend into chaos. Yet another scrap from the intellectual dump of context-free quotations is the statement "thoughts without content are empty, intuitions without concepts are blind" (*CPR* A51/B76).[6] But it is in his tacit critique of the empiricists' underdeveloped conception of experience that Kant really zeroes in on the dump: "Since, however, if representations reproduced one another without distinction, just as they fell together, there would in turn be no determinate connection but merely unruly heaps of them [*bloß regelose Haufen derselben*], and no cognition at all would arise, their reproduction must thus have a rule in accordance with which a representation enters into combination in the imagination with one representation rather than with any others" (*CPR* A121).

Heaps of representations are disorderly, commemorating in their static condition the dynamics of the fall, having plunged as an indistinct mass, together but without any "determinate connections." Evocative of Plato's assessment of vision unaccompanied by *noēsis*, Kant's mental dump nevertheless diverges from that assessment in several respects. Three of these are worth highlighting:

1. The impressions reproduced "just as they fell together" do not, according to Kant, add up to an experience; in Plato, seeing yields a vague experience of magnitude.

2. The principle that could bind the sensory heap together is a synthetic rule of combination (*Verbindung*), rather than the separations of Platonic discernment.

3. The massive hordes (*Haufen*) of impressions are a plurality, albeit, in keeping with Kant's numeric categories, a plurality that has not yet appeared under the aspect of unity in a totality. In Plato's sight, magnitude is uncertain: the great is seen as small, and the small as great.

Still, both philosophers fight against a common enemy, the fusion and the confusion of the many dumped in the guise of an undifferentiated, yet fractal, mass.

It bothers Kant that numbers and the category of relation are unsuitable to the unruly heaps of impressions. If any one impression could be combined with any other, he conjectures, then imagination and the acts of meaning-making would come to naught. The rule has to do, primarily, with exceptions from the general associability of everything with everything, that is to say, from the dump, extending all the way to strands of ecological thought with their slogan "everything is interconnected," which is lethal to imagination. Determinate connections are none other than relations eventuated from selective associations, "with one representation rather than with any others." Kant's word for this necessary cherry-picking is *affinity*, "the *affinity* of the manifold [*die* Affinität *des Mannigfaltigen*]" (*CPR* A113).

At variance with the heap, where distances between the confused parts have been both eliminated and exponentially increased, affinity is proximity respectful toward the borders, edges, or ends of the terms it articulates (from the Latin *ad+finis*: "to the end," "to the border"). The randomly heaped is not interrelated because the borders between the heap's sundry components are nonexistent, borders that, though tenuous, porous, permeable, and open to transgression, should be in place as so many breathable membranes. The denouement of limit-effacement is not freedom but the vilest despotism of the dump over whatever and whomever falls into it without affinity and, therefore, without a chance for solidarity among the dumped. The dump is not given in Kantian experience, on which it bears down, transcendentally, as a negative condition of possibility. Boundless, it marks off the limits of cognition no less effectively than the thing-in-itself. Perhaps, the thing-in-itself *is* a dump. We will never know. What we do know is that things-for-us are rapidly turning into dumps, for us and in themselves.

Aristotle thinks otherwise, presaging James's view. The first mode in which the world is given to us is a confused pile, a muddle of *ta sugkechumena*. He begins *Physics* (the first chapter of the first book) with a mention of first givenness: "For us, what is manifestly and clearly first is rather the confused [*Esti d' hēmin prōton dēla kai saphē ta sugkechumena mallon*]" (184a, 22). The irony built into the phrase should not be lost on us: "what is manifestly and clearly first" is the covered-over, the withdrawn, the nonmanifest, unclear, elusive. The world initially appears as a dump.

Another set of questions comes first, preceding the inquiry into modes of givenness. Who is this *us* in the Aristotelian *for us*, to whom the entire thing is manifest and clear? Must "we" revisit the first beginning in an inquiry about nature—*peri phuseōs epistēmēs* (184a, 15)—so as to overview the dump of the primarily given from a detached and impartial perspective of the whole, *ta katholou* (184a, 26)? Does such an inquiry announce the coming to a close of the first beginning? Must "we" subtract ourselves from the confusion of the heap and, moreover, find ourselves, arriving at the Aristotelian *for us*, in the act of this subtraction? If so, then by what means to achieve this feat now, when the dump has become global, has englobed every perspective and vetoed any other mode of givenness?

To continue: Aristotle begins at a beginning that threatens to wind up in a dead end of confusion. For us, living and dying on a global dump, that vague threat has materialized. The beginning is immediately, without transition, the end. For us, the Aristotelian knowers, knowing begins with the confusion whither the philosopher comes round again and passes his judgment on the mess: *clearly unclear!* Having barely begun, the cognitive endeavor is not meant to terminate in a mental babel, assuming that thinking catches up with *phusis* around which or about which (*peri*) the inquiry roves, tracking its perimeter. Aristotle is resolute in his belief that the dump is not in matter; it is, he insists, in our immature approach to *what is*. But if the first beginning takes root, endures, and pulverizes all further unfolding, as it has already done, then matter, the elements, and the world are no less of a dump than is incipient knowing. Correspondence theory of truth has its field day: the dump inside matches the one outside, and the difference between the inner and the outer also succumbs to the dump's leveling force.

In the abiding immaturity of sundry confusions, our lot is the second, Shakespearean childishness shorn of the Christian overtones of the second innocence (Matt. 18:3). We are the grey-haired and shriveled infants, with not a drop of the green freshness (*viriditas*), which Christian mystic and saint, Hildegard of Bingen, associates with the self-renewing power of creation. Paralyzed in the first beginning, we would repeat (not knowing it as a repetition, not quite knowing what it is that we are doing) the experience of little children (*paidía*), who "first [*próton*] call all men fathers and all women mothers, and only later [*husteron*] differentiate among them" (184b, 13–14). The dump is where the closest relations are the same as the furthest, all treated with the same shoulder shrug. One is indifferent in the first place, in the nonplace of the first, to oneself as someone who has issued from *this* mother and *this* father. The fruit of generic progenitors is a generic self. The dumped one is not an I. The piling on of relations unbinds them, revokes the possibility of individuation for oneself and for the jumbled contents of the given.

As for Hegel's dump, it is the desert of sense-certainty. Before he displays its bankruptcy masquerading as infinite wealth, Hegel launches an appeal not to change anything in the immediacy of the beginning. "Knowledge, which is from the first or immediately our object [*Das Wissen, welches zuerst oder unmittelbar unser Gegenstand ist*]," he writes, "can be nothing but immediate knowledge [*unmittelbares Wissen*]. . . . We must also behave towards it *immediately* or *receptively* [unmittelbar *oder* aufnehmend] and to alter nothing in it" (*PhG* §90).[7] Easier said than done! "Immediate" knowledge is amnesiac as to the way it has come to know, unaware of the stages it has gone through and the hurdles it has jumped. Yet, to have knowledge for an object is to have already mediated it, also with the participation of self-consciousness. We, who know knowledge itself, are not we, who live in the immediacy of immediate knowing: the rub, the frictions impelling dialectics are palpable in this divergence between us and ourselves. In order to receive (*aufnehmen*) immediate knowledge, we must be receptive to ourselves as we were at the beginning and, more than that, "alter nothing in it," in the beginning and in that first *we*.

Reenter the dump (which, in *The Phenomenology* and in previous philosophical treatises starting with Aristotle's *Physics*, is a rewind-and-replay of Plato's cave), reenter it—Hegel implores—but tread lightly and do not touch anything with the magic wand of thought, as if your return has not taken place and as if you are there for the first time ever. How lightly *can* we tread and on what in the midst of a massive and sudden fall? Wouldn't behaving immediately toward immediacy, across the scission between us and ourselves, mean that nothing would ensue? That the later *we* would be dumped into the earlier? Hegel is the first to disregard his own advice as he forges relations in, mediates, and objectifies the dump's not-yet, or already-not, things and thoughts.

Although a return to the cave keeps replaying itself, returning over and over again, it does so according to different scripts. Do we get a good overview of the cave's world or unworld from above, at the feigned remove of *topos ouranios*? Or do we attempt to disentangle it from within while claiming that no meddling is taking place, that the philosopher is merely following the thread of the jumble's own self-disentanglement? Hegel desperately wants the later *we* to follow the earlier under the pretense that they are one and the same, the distance between them nil. He indulges in wishful thinking here: the immediate object is mediated, if only by its status as object, which ejects us out of the dump, where we think we are. Effectively, we are outside it, whenever we think that we are within. But, then again, it could well be that all thinking is essentially wishful.

That said, the Hegelian dump is a pure, unmediated abstraction of *this*, *now*, and *here* flipping in a blink of an eye into *that*, *then*, and *there* (*PhG* §§95–108).[8] Sense-certainty appears to yield "infinite wealth," *unendlichem Reichtum*, which, as soon as we take hold of it, presents itself as boundless penury ("no boundaries are to be found for it [*keine Grenze zu finden ist*]") (*PhG* §91). With its sudden reversals, of the kind that upend plenitude and turn it into almost nothing, sense-certainty is the missing link between the dump and the desert. I, *this* I here-and-now (*hineni!*), am dumped together with the rest of the prima facie singular but, in fact, generic, neither particular nor universal, placeholders for being. Infinite wealth is boundless penury, since *this*, *now*, and *here* are uninvolved with the entities that pass through them and move on without stopping to another formal *this*, another *now*, another *here*. Unimpeded freedom of movement is actually paralysis in the virtuality of formal abstraction. Sense-certainty converts time and space themselves into dumps of interchangeable instants and places, commensurate in their overall insignificance.

Inchoate abstraction is a heap unshaped by the work of determinate negation. Nowhere is this more glaring than in the case of numbers purporting to express lawful relations among properties. In the role of the principal scientific tools for understanding reality, numbers are piled "in this heap," *in diesem Haufen*, to which they reduce the reality in question (*PhG* §290). At issue is not quantity as such but the concept-free abstraction that directly (once again: immediately) translates qualities into numeric values, forsaking the arduous dialectic of quantity and quality. A formula is a pile of variables, presupposing that "properties, as *existing* [*als* seiende], are just lying there and are then taken up" into it. The formula's inflexible mathematical lawfulness "demonstrates the abolition of all lawfulness [*die Vertilgung aller Gesetzmäßigkeit darzustellen*]" (*PhG* §290). Alas, the rigid arbitrariness of notation severed from existence *is* our existence, a digital dump of *1*s and *0*s.

our polluted senses

Little has changed since Hegel's diagnosis with the introduction of Big Data, save for the scope of "the abolition of all lawfulness."[1] Increasing by the minute, falling on us and with us, the information dump lets one category—quantity—override the rest. Set over and against the actual properties of things, it throws a challenge to existence. In our implacable blooming, buzzing confusion, we are assailed not so much by our "eyes, ears, nose, skin, and entrails at once," as by formless information poured onto them from whichever direction. Between their upper and lower thresholds, the senses buckle under overpowering streams of data. Sense-uncertainty, disorientation experienced in the information dump, replicates and aggravates the effects of Hegelian sense-certainty that spins the world of interchangeable *this*es, *here*s, and *now*s, aloof to their transient instantiations.

The senses are also under an unremitting attack by the sensory stimuli themselves. Bemoaning light and sound pollution deeply engrained in the fabric of advanced urbanism is a commonplace. The situation has got so bad that the state of Idaho in the United States has decided to create a dark sky reserve in order to preserve its remarkable conditions that make the interstellar dust clouds of the Milky Way visible on a clear night.[2] Less frequently we deem our senses themselves tainted, though *light pollution* is in fact a euphemism for the enervation of human, animal, and plant *vision*. For, ultimately, the problem rests on the receiving end of the phenomenon that contracts the sphere of what is phenomenally accessible.

The exteriorization of pollution, its relegation to light rather than the eye or photosensitive cells and tissues, is meant to reassure us that we are the islands of inner purity in an ocean of environmental contamination. Forgetting the ancient wisdom that espied the macrocosm of elemental fire in the microcosm of the eye's inner luminosity, we carry on as though the blotting out of the starry sky did no harm to seeing and thinking. In reality, the formulation *light and sound pollution* could not be more misleading. It is the senses that are desensitized to subtler cues by the intense stimulation they receive or, more precisely, fail to receive. City dwellers can grow so accustomed to this state of affairs that they will no longer notice it, nor everything it retracts from the field of perception, anymore: perceptual thresholds shift upward as radiant energy and strong vibrations are dumped on sense organs nonstop. Just as the foulest parts of the global dump dodge decomposition and do not smell of rotting, and just as the dump's companion affect is the apathetic *whatever* . . ., not the sentiment of horror and shock, so the most disturbing quality of its impact on sensation is imperceptibility, not insufferable hyperstimulation.

Passing similarly unremarked is the pollution of the sensorium beyond its visual and auditory registers. The way the glow of bright city lights causes twinkling stars to recede from sight is comparable to sugary and salty foods foiling the palate's appreciation of the more delicate flavors. The overpowering scents wafting from perfumes or candles induce the same reaction in the olfactory system. It could well be that the tactile sense, which philosophers berate for its material anchoring in comparison to the distance senses of vision and hearing, is the last bastion of differentiation. That said, in touching, too, we have drastically narrowed down the field of what can or should be touched. As we spend much of our time caressing the smooth and glassy touchscreens of "smart" phones and tablets, it remains to be seen, or touched upon, what this new habit might do to tactility.

The sensory dump is a desert, the one we harbor within. Our peculiar dilemma is that of impoverishment through surplus. In Plato's sun analogy, the source of visibility was a generous excess that offered light and life, luminosity and warmth (in a word, incandescence). Plato recognized that it was impossible to look directly at the wellspring of vision, which retreated from the sense it activated. *Our* sun is the radiance of the earthly, not the heavenly, city that, itself visible, plunges the shimmer of celestial bodies into invisibility. Salt and sugar are the flavors *du jour* (which is to say: *de la nuit*) of the postmetaphysical sun, their assault on the palate masking other tastes. They are the essential and toxic[3] ingredients of "junk food," an expression that corroborates the general logic of ingested pollution. More and more, the stuff of our senses is detritus when too much of something—of one thing—implies too little of everything else, curtailing the range of what our bodies meaningfully receive from the outside. Our receptor cells are becoming garbage receptacles, by now crammed full of sensory trash.

We have managed to turn the senses against themselves by pitting a light against lights, a sound against sounds, a flavor against flavors, an aroma against aromas. The tendency is toward a blatant simplification in the field of possible experiences owing to the eclipse of multiple stimuli by one or two that outshine, outsmell, etc. the rest. We live in the state of *sensory underload* when dominant sensations muscle out those that lay a weaker claim on our attention. The massive fall of a stimulus rarefies the senses and, making them abstract, voids their own discernments. In this way, the dump produces the senses as the facsimiles of a disembodied mind, even if, in this capacity, they will never live up to the expectations of conceptual thought they are prompted to emulate. The dice loaded against them, the senses are subject to further devaluation and abuse.

Rather than input sites for information to be processed by our minds, the senses are the crux of our embodiment, the synergistic interfaces of consciousness and the world that *give rise to* consciousness and the world, which is not the same thing as reality. As Merleau-Ponty has it: "The sensible is what is apprehended *with* the senses, but now we know that this 'with' is not merely instrumental, that the sensory apparatus is not a conductor, that even on the periphery the physiological impression is involved in relations formerly considered central."[4] The erosion of the sensible erodes who we are, not (only) what we come to possess. The squashing of the senses by the stimuli dumped onto them is the quashing of our being. A "merely instrumental" interpretation of apprehending the sensible *with* the senses (prior to its instrumentalization, this *with* betokens the synergy of primordial sociality, being-with-the-world) is an early warning sign of ontological pauperization. Despite retaining the *with*, the new logic of the senses takes it to mean *by means of*, denies the preposition's articulatory effects, and dissociates consciousness from the world. Disarticulation is the noxious energy of the dump that weighs on, penetrates, and wreaks havoc in the sentient body.

The receptivity of our sensorium to the finest discernments and to massive bombardment by the crudest stimuli is the poisoned gift of existence, the *pharmakon* of psychic life. Blaming modernity or capitalism is futile; the potentialities of embodiment themselves mold the senses into (potential) receptacles, the trash-bins of experience. Light, flavor, sound, scent, and touch pollution result from tremendous discharges of a stimulus exclusive of others and dumped upon all those in its vicinity. Obesity that is down to a regular consumption of junk food is a conspicuous cellular and tissue-based archive of the dump, with fat deposits supplying "objective" evidence for the ingestion of an impoverishing surplus.

With rare exceptions (e.g., the proposed dark sky reserve in Idaho, itself linked to plans to promote tourism in the state) pragmatic and functional concerns outweigh worries about the ontological facets of sense ecology. Granted, the illnesses noise and light pollution induce, from insomnia and depression to hypertension and ischemic disease, impair our physical and psychological well-being. But the aesthetic damage they inflict is irreducible to pedantic, aestheticist laments about the ugliness of mass-produced and mass-consumed material reality. The dump takes charge of "the distribution of the sensible" (Rancière), trawling the previously visible into invisibility, the previously audible into inaudibility, and so on. The pollution of the senses imprisons the body in itself, with overpowering stimuli for jail fences. At the extreme, it robs the subject of her world. On the outward side, the fences it erects are wedged between the body and a body, sentient flesh and a cadaver. Inwardly, they redraw cognitive and perceptual maps, promoting a glaring simplification in the course of what purports to be the age of complexity.

Seeing that the aesthetic domain is incomparably broader than aestheticism would admit, individual reactions to the unbeing that besieges and indwells us will be of little consequence. Some among us might be sufficiently privileged and wealthy to seek private escape routes from the inner and outer dump, be it in the quiet of meditation classes or the pleasures of gourmet dining. These niche solutions constitute an upscale market of experiences at a time when the material form of experience has been decimated. They sell a lie, merely anesthetizing their buyers to the operations of the dump. Pushkin had an apt designation for it: a feast at the time of the plague. "As we lock ourselves indoors when the prankster Winter comes, / So we will do when the Plague approaches! / Candles we'll light and wine pour, /

Merrily drowning our minds in it. / And, throwing feasts and balls, / We will glorify the kingdom of the Plague."[5]

In the oblivion of our inebriated minds, we forget that what is being dumped, and dumped upon, is being itself. The dumped is *what is*, all of it. Of course, being cannot be seen, touched, heard, smelled, or tasted; being and radioactivity are (what a terrifying thought!) akin in this respect, though ionizing radiation can be measured using special instruments, such as the Geiger counter, while being cannot. But the senses are, up to a point, our guides to ontological domains, a little like Virgil whom Dante follows through the circles of hell and the purgatory, or Beatrice who leads him through paradise. Can we rely on their guidance today, when radiation, microscopic water contamination, and airborne toxins elude the sensory register? On the one hand, marked out and steered by the senses, the path to or through being ends abruptly in the dump with its disorientation and lack of discernment. On the other hand, not insightfully following the senses is walking straight into the dump. Our polluted senses signal that the two hands (on this clock of the world; on this world-clock) are one and the same. The clock strikes midnight.

toxicity

The dump is toxic. Were it still possible to isolate its strata, we could enumerate the chemicals that harm living bodies and the elements, classifying them apart from a polluted sensorium, venomous imagination, and virulent intellection. In reality, these levels are intermixed: noxious thoughts and poisoned senses, toxic built environments, social milieus, and contaminated ecosystems merge and reinforce one another.

Flying from every direction, the arrows of toxicity do not discriminate among those they hit in a "toxic flood," the anthropogenic emission into the environment of over 250 billion tones of chemicals a year.[1] Like the generic stain of the original sin, they do not single their victims out by means of a negative and lethal individuation, an alien intentionality that would aim at *me* as in the case of an animal who, to defend itself, releases poison toward a threatening target. Toxic materials massively fall on and pervade whatever and whomever crosses their innumerable paths. They hit "me" as a lump of flesh not at all distinct from the flesh of a rodent, a cockroach, microbes, or a dandelion.

As I undergo traumatic deindividuation, toxicity fleshes itself out. It gives itself a body in my body and a world in every elemental region. Its darts and missiles come from the inside, too: from my toxic corporeal and psychic interiority—for instance, the desire to cleanse my garden of unwanted intruders so as to have a perfect lawn. In toxic social and political environments, harassment and persecution proceed along similar lines. The victim is not individuated by victimization; despite varying fetishes, predatory predilections, and degrees of unwanted advances, sexual infringement on women precipitated by "toxic masculinity" is indiscriminate, as is the reaction to it emanating from #metoo. We are all dumped by patriarchy's toxic order, men and women alike.[2] Toxicity is our unmoved mover, an arrow bent into a circle.

Besides blue-green algae that poisoned the atmosphere with the oxygen they exhaled for millions of years until wiping themselves out, there is probably no other creature in existence more adept at poisoning itself and its lifeworld than the human. So much so that poison now organizes or disorganizes, disorganizes in organizing, both the poisoner and the poisoned. Having surpassed a manageable quantitative threshold, toxic destruction has attained the qualitative power to create mutilated bodies and worlds. The toxicity of the dump is a ragbag of chemical-laced water, soil, and air; disordered reproductive and endocrine systems; aspirations to infinite growth without decay; energy dreams that bequeath to countless generations to come depleted (what a misnomer!) uranium sometimes recycled (another misnomer) in munitions;[3] vision debilitated by light pollution; skyrocketing cancer rates, or else a proliferation of cells that refuse to die; pesticide- and insecticide-imbued pastures and forage crops. But the red thread of the dump's mangled components is *ontological toxicity*—that which does not pass and, in not passing, warrants the annihilation, the rapid passing away, of all else. Theological longings for life everlasting, infinite market expansionism, metaphysical constructions of a true unchangeable reality, oncological disease, and radioactive waste are ontologically toxic. At their core is the kind of being that, protecting itself from nothing and eschewing becoming, lapses into the very thing it is so scared of. The being that in its absolute isolation is unbeing, traditionally labeled *evil*.[4]

Take the desire for immortality, which envisions individuation without finitude, or life without death—disjunctions as nonsensical as neoliberal growth without decay. Is the secular iteration of this desire not premised on the conviction that, were it to be fulfilled, I (or, at best, I and those closest to me) would be rescued from the clutches of death, while the entire world perished? Certainly, the underlying conception of who or what *I* am is paramount here. If you are convinced that this body you call yours is of the essence, then cryopreservation is a way to act on the desire to be immortal. If consciousness matters most, then it is imperative to save it on a durable substratum, to upload its data onto a supercomputer or some such. Both solutions posit a practical separation of the mind from the body and of the I from the world, so that the former participant in each would-be relation could outlive the latter.

The religious horizon for eternal life was a perfect community of other righteous souls reconstituted in the yonder of heaven. Those languishing in hell were isolated from each other (and from god) by their horrific punishments and sufferings. With this historical context in mind, the secular vision of immortality saves, in the guise of its ideal, the hellish image of an alienated individual, cut off even from certain facets of itself. Ontologically toxic, it hampers the passing of a given body or mind and sanctions the destruction of their disposable existential *wherein*—the world and the body, respectively.

One's reluctance to pass away "for good" is included in the logistics of the dump where heaps of isolated debris shun rotting. As does a cancerous growth, in which a group of cells rebels against finitude, maintains itself intact past its due, multiplies the quicker the less differentiated it is, invades other tissues and organs, and leads the organism to its demise. In aggressive tumors, the loss of cellular structure and function, massive cell division, undifferentiation, and metastatic extension to other parts of the body are the clone characteristics of the dump. Truth be told, before they spread beyond their original site, malignant growths are the metastases of the dump in the oncological patient's body. Cancer is a privileged physical, physiological vehicle for ontological toxicity. The disease gives birth to death and empties actual being into unbeing by instantiating immutable (resistant to destruction) and at the same time highly mobile, volatile (metastatic) existence in biology.

Toxic substances are dumped into rivers and lakes, the atmosphere and the soil. Their repercussions are also dump-like, whether they contribute to the "global cancer epidemic"[5] or indiscriminately contaminate the organisms that imbibe them through their membranes. In her pioneering book *Silent Spring*, Rachel Carson refutes the argument that herbicides should be the weapons of choice in a *targeted killing* of unwanted plants. Although the toxicity of these substances is ramified according to varied biochemical, physiological, genetic, and metabolic scenarios, their effects do not comply with the nominalist boundaries of natural classification systems: "The legend that the herbicides are toxic only to plants and so pose no threat to animal life has been widely disseminated, but unfortunately it is not true. The plant killers include a large variety of chemicals that act on animal tissue as well as on vegetation. [. . .] The herbicides, then, like the insecticides, include some very dangerous chemicals, and their careless use in the belief that they are 'safe' can have disastrous results."[6]

The careless use Carson decries depends on a projection of the alienated *I*, severed from the environment, onto the world at large and its inhabitants. The credo buttressing thoughtless and insensitive use is that the intended targets of herbicide's toxic arrows are self-contained; that only that which is unwanted is caught in its crosshairs; that harmful chemicals do not bind to and poison also the populations of cultivated plants; that their impact beyond the flora is negligible; that their aerial spraying, misting, or spreading by means of rope wick applicators or blanket wipers does not contaminate the air, the water, and the soil. Carelessness is the practical and psychological reverberation of the indifference ruling the day and the night (that is, the nocturnal day) in the dump.

Reliance on toxins in the hope of controlling environmental processes and interactions, ensuring cleanliness, or regulating agricultural production is itself uncontrollable. The term *toxic flood* transmits the gist of this uncontrollability by singling out one element, water, gushing with irrepressible force. The cleansing power of the aquatic element, symbolizing religious and physical purity,[7] all but evaporates: water is no longer the milieu wherein every evil and impurity may be diluted, but liquid excess suffused with the immensity of the problem. The earth is reaching its limit to churn out from putrefaction and "distemper'd corpses" the "sweet things" Walt Whitman marveled at in his 1856 poem "Compost." It is, instead, smothered by those non-decomposable things that, while eliciting no disgust, disrupt every transition and transformation from decay to a new growth. More precisely, the extant version of *every* element is a dump for toxic materials: an aerodump filled with smog; a hydrodump impregnated with runoff, sewage, and plastics, including those still frozen in the rapidly melting Arctic ice; a pyrodump of global warming; and a geodump with growing deserts, spent nuclear fuel storage facilities, industrial traces in geological strata, and heaps of debris. The geological era of the Anthropocene (*geology* itself may become anachronistic, once the *logos* of the earth has been bulldozed into the geodump) is but a drop in the sea of elemental metamorphoses into the world, or the unworld, of the dump that toxicity makes-unmakes.

Geology without *logos* should, despite everything, have a special place in our age's self-understanding. Before the toxic flood (where is

Noah's ark in it?), the earth was a synecdoche of the fourfold, an element that stood for all the elements. After the frenzied unleashing of toxins and carbon emissions, the earth onto which everything falls continues to serve as a model for pyro-, hydro-, and aerodumps. Satisfying an old metaphysical yearning, all is materially becoming one and the same. Even as we disconnect from the earth—whether it refers to agricultural soil, the land, or the planet—the elements are earthified in a garish substantiation of the link that exists in English between pollution and *soiling*. (*Landfill*, too, is a chillingly trenchant word, insinuating that land is desolate and barren, not yet full, before its impregnation with trash and with metaphysical garbage.) The elements converge on infertile silt, fecundating nothing but death: dense with the charred, polymerized, and polycondensed remnants of fossils from vast underground deposits; heavy metals; nitrogen and phosphorus; nitrates and plain garbage.

Gaia, the Greek for *earth*, also connoted a certain density in a region supportive of human dwelling and prepared to receive the dead.[8] Earthly opaqueness thrust onto the elements harks back to the antiphenomenality of the Heraclitean *kechumenon*, haphazardly poured out, piled at random, and blanketed over. Obscurity reigns where it does not belong: in the transparent abysses of water and in the expanses of air that no longer lets the light of the sun or the stars pass through, trapping heat in the atmosphere and tearing asunder the two powers of fire, the luminous and the thermal. Ontologically toxic, the elemental dump ousts the elements from their proper regions, unfastens them from each other and each from itself. Last but not least, the dump is—in contrast to the earth that has imparted opaqueness to water, air, and fire—too unstable to support anything. Yet, it does eagerly take in the dead.

Figure 6 June 9, 2018, Cruz Quebrada, Carbon black level (PM2.5): 6.5 µg/m, Carbon ink print, 100 × 66 cm, Anaïs Tondeur 2018–20

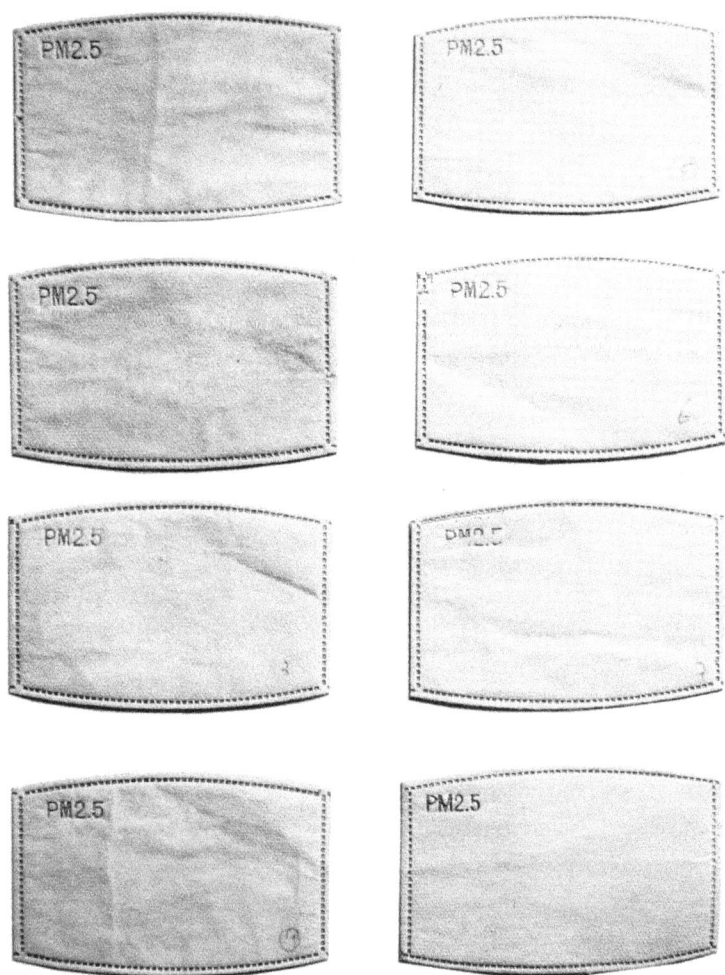

Figure 7 Anti-PM2.5 masks, Carbon Black, Anaïs Tondeur, 2018–20

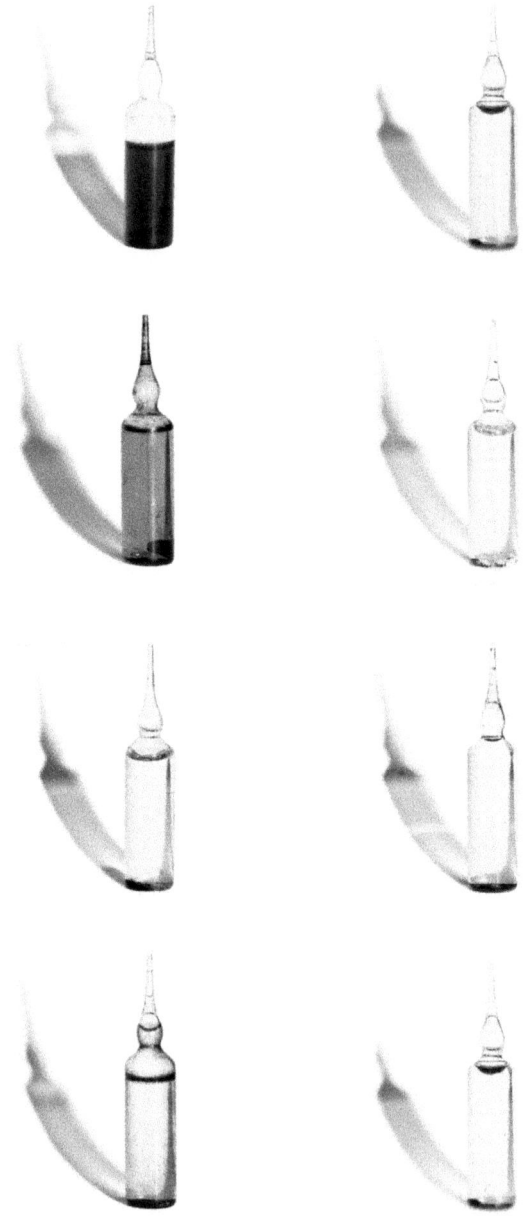

Figure 8 Extracted carbon black particles, Carbon Black, Anaïs Tondeur, 2018–20

Figure 9 June 11, 2018, Cruz Quebrada, Carbon black level (PM2.5): 7,2 µg/m, Carbon ink print, 100 × 66 cm, Anaïs Tondeur 2018–20

Figure 10 June 10, 2018, Cruz Quebrada, Carbon black level (PM2.5): 12,3 µg/m, Carbon ink print, 100 × 66 cm, Anaïs Tondeur 2018–20

Figure 11 June 15, 2018, Cruz Quebrada, Carbon black level (PM2.5): 9,1 µg/m, Carbon ink print, 100 × 66 cm, Anaïs Tondeur 2018–20

Figure 12 June 13, 2018, Cruz Quebrada, Carbon black level (PM2.5): 9,1 µg/m, Carbon ink print, 100 × 66 cm, Anaïs Tondeur 2018–20

Figure 13 June 12, 2018, Cruz Quebrada, Carbon black level (PM2.5): 5,4 µg/m, Carbon ink print, 100 × 66 cm, Anaïs Tondeur 2018–20

Figure 14 June 20, 2018, Cruz Quebrada, Carbon black level (PM2.5): 7,9 µg/m, Carbon ink print, 100 × 66 cm, Anaïs Tondeur 2018–20

Figure 15 June 21, 2018, Cruz Quebrada, Carbon black level (PM2.5): 2,1 µg/m, Carbon ink print, 100 × 66 cm, Anaïs Tondeur 2018–20

shitty apocalypse, or scatological eschatology

There is another, decidedly unpoetic, vulgar sense to dumping: an act of defecation. Semantically and sonorously unrefined, the expression involves "having" or "taking" a dump. An escalating confusion: verbs that express a voluminous release, passing the contents of the bowels out of the body, are those that normally denote retention (*to have*, *to take*). Giving is taking here, and for a good reason: the massive discharge from one's innards fertilizes neither the soil of nature nor that of culture; it neither nourishes a finite, sustained, and sustainable growth nor makes room for the future. Instead, it transfers the stuff that has been clogging the inside outward and clutters the world. To "take" a dump is to externalize that which causes indigestion, to exacerbate the constipation of life, to steal space and time by occupying them with loads of undecomposable detritus. It is to give what forecloses all future giving and the giving *of* the future.

And, in fact, our eschatologies are overwhelmingly scatological. Terrified of drowning in our own shit, we realize that the brilliant apparitions and the earth-shattering sounds of the seven trumpets announcing the end of the world and salvation for the righteous are figments of an outmoded, naïve imagination. It dawns on us, in those instances when repression is too weak to contain the thought, that the approaching inglorious apocalypse will more or less quietly and imperceptibly trash the entire planet. The shrouding of the earth in pollution and of its atmosphere in CO_2 emissions reverses the original meaning of the apocalypse as the final uncovering of reality, the truth revealed at the end of times. What the actual unveiling lays bare is a deadly, stifling veil thrown onto the elements.

But why adopt the language of defecation in the context of the environmental crisis? For two reasons.

First, with regard to the body's physiological functions, a growing global population means that the mass of human excreta is also increasing. Nearly one billion people "still defecate in the open, for example in street gutters, behind bushes or into open bodies of water" and "at least ten percent of the world's population is thought to consume food irrigated by wastewater."[1] After processing by sewage treatment plants, effluent wastewaters carry microplastics and anthropogenic nonsynthetic fibers aplenty into the sea.[2] Numerous pharmaceuticals, notably sulfamethoxazole, cannot be removed through conventional sewage treatment and, as a result, are dumped into the environment.[3]

Second, because human life is not only biological but also cultural, economic, and technophysical, there is a surfeit of its residua over and above the urine, feces, and carbon dioxide our bodies secrete, excrete, and exhale. Our works, the objective outcomes of our exertions, are the things we give off and live on, evocative of a tree that bursts forth in leaves and fruit, lets them fall and decompose around the roots, and feeds on their decay. Prosthetic bodies produce prosthetic feces.[4] This is not a crass metaphor, but a statement of technological metabolism that ends in a disruption of metabolic logic.

Augustine approaches the issue from a diametrically opposed angle and excels in a figurative construal of fruit in terms of human works. According to his take on the creation story in Genesis, the "fruit of the earth are to be allegorically interpreted as meaning works of mercy, which are offered for the necessities of life from the fruit-bearing earth" (*Confessions* XIII.26.38). Remembered by god, these fruit are kept forever in his infinite stomach. As usual, Aristotle is not far from Augustine's mind here. In *Parts of Animals* the Greek philosopher puzzles over the fact that plants have no equivalent to the intestine in animals: they are "without any part for the discharge of waste residue." "For the food which they absorb from the ground," he adds, "is already concocted, and they give off as its equivalent their seeds and fruit" (*De Partibus Animalium*, 655b). Plant roots function as subterranean mouths, and fruit correspond to the leftovers of the nutritive process, to excrement. (Hildegard of Bingen, whom we've already briefly met on these pages, is more discerning in this respect: only the juices of "useless herbs that should not be eaten" are "comparable to human waste [*egestioni hominis comparantur*]" [*Physica* I.6].) Aristotle thus contracts *tō threptikon* to a single capacity for nourishment, where reproductive mechanisms (seeds and fruit) are the waste left over from nutrition.

Human works, the fruit of our labor and industries, are the droppings we leave behind. But what if the *behind* is also ahead of us, above and below, right and left? Overlooking aspects of vegetal life in our existence (for instance, the cultural soil into which works and texts decompose only to promote future thinking at the price of their self-contained identity, as Derrida elegantly puts it[5]), we behave like animals who run away from their excrements, not like plants recycling the organs they shed. Our separation from the places we inhabit undoubtedly exacerbates animal locomotion, such that every *here* is perceived as a *there* still before we hastily pass through it. The strange case of *Socratea exorrhiza* also known as the Walking Palm apart, plants have no opportunities to abandon the habitats wherein they grow; they must live with and live off the physical by-products of their existence. Human vegetality is perhaps most readily appreciable on the planetary scale, once the soil, the oceans, and the atmosphere get filled to the brim with our shit and there is nowhere else to run. And yet . . . On the brink of a vegetal planetary consciousness, the scale changes: the ideological apparatus siphons imagination, energy, and attention off to the possibility of an interplanetary species-life with permanent colonies outside the earth. In this storyline, just as in its theological prequel, salvation is reserved for the select few, the ultrawealthy elites who envision the continuation of life in a cosmic elsewhere. Without giving it a second thought, they are prepared to abandon the earth, a pack of animals on an interplanetary run from their excrement.

Fruit and works, excrements and superfluities, fall from trees, animals, and humans. It is a matter of *letting fall*. But dumping is *making fall*, its force felt in the massiveness of the falling and the fallen, of release and impact. Rather than culturally, economically, or physiologically digest what is needed for existence, the global dump reprocesses living organisms, the elements, and the world itself into loads of shit. On the one hand, the deranged metabolism that has automatically taken the place left vacant by the old subject-object of history gathers pace and releases everything and everyone from its guts, making them plunge to their demise. On the other, it slows down to a crawl when that which has dropped down and out both eschews decomposition and does not retain its past identity. Piling high, the dumped fragments of the excreted world defy time. Radioactive or not, the waste that falls this way signifies one thing and one thing only: a fallout.

The concomitance of acceleration and slowdown is a token of ontological toxicity in the closure of metaphysics. The ambiguity of this closure, which does not come to a close, is that, as I have already noted, excreting metaphysics, the age of the global dump externalizes into material existence the indigestion and the indigestibility of metaphysical (unchangeable, eternal, resistant to metabolism) being. We are undergoing a period of colossal displacement: metaphysical indigestibility moves from the innards of thought to the world in the guise of products that obliquely and diabolically realize the dreams of immutability, of liberation from temporal inexorability. Our actuality takes a dump with the ontology *of* metaphysics mixed with everything indigestible *for* metaphysics—nonidentity, the other, absolute exteriority, infinite difference. Cleansing is cut from the same cloth as pollution: on a practical level, the chemicals used to clean our homes and offices poison the earth;[6] on a theoretical level, the foremost factor in environmental contamination is squeaky clean, pure being isolated from nothing and immune to the metamorphoses of becoming. It is in the name of what is supposed to remain eternally virginal, untouchable, and incorruptible that the world is converted into a litter box. In turn, a sunny future outlook is possible thanks to the presumed resilience and phoenix-like rebirth of finite existence from the flames of devastation.

Although the approaching (unless it has already happened) shitty apocalypse stirs up free-floating anxiety and dread, it is also a clandestine source of pleasure. Psychoanalysis is critical for understanding the paradoxes of the complex affective reaction it elicits, while a quick glance at the mushrooming academic discussions of the Anthropocene provide us with ample diagnostic materials. The discourses about the Anthropocene combine repulsion from and attraction to the planetary excrements of economic development, shame and pride enkindled by the vestiges of industrial activities stored in the earth's strata. The disavowed positive aspects of our Anthropocenomania closely resemble a child's enthrallment with the works of her bowels that she views as her "gift" to the world, and, later, as her "babies" (SE 7:186).[7] The decision to bestow the name *Anthropocene* onto the era of a significant anthropogenic impact on the earth and its atmosphere is symptomatic of a projection: "I am human (*anthropos*) and, therefore, this is also *my* age; I have also had a hand, or another body part, in it." (The obverse—dissociation—is the lot of those who reckon themselves posthuman: "I am no longer human and, therefore, this is *not* my age; I have nothing to do with it; I am an innocent victim of the hands, or other body parts, pertaining to the despicable and now obsolete humans.") In the Anthropocene, *anthropos* spots and recognizes itself in a grimy mirror of geological shit. Gone is the permanent and nagging indeterminacy of the human experienced as a troubling open question in twentieth-century existentialism. Instead of the question without an answer, we turn ourselves into the answer without a question. The meaning of humanity is objectively and scientifically fixed by a cross-section of the geodump, with which *we*, everyone living-dying in today's long night, are urged to identify, in which we are urged to identify ourselves.[8] Such an identification, in turn, demands a serious libidinal investment, cathecting the freshly minted subject to the unwieldy dumped object (planet-changing pollution, archived in geological strata). And cathexis, as we know, is never unreservedly negative. Consciously assuming responsibility and enduring the self-flagellations of the superego may be no more than a pretext for unconsciously joining in the pleasure of collective achievement.

The psychic mechanism of anal gratification Freud has examined in "Three Essays on Sexuality" sheds additional light on scatological eschatology.[9] "Children who are making use of the susceptibility to erotogenic stimulation of the anal zone betray themselves by holding back their stool till its accumulation brings about violent muscular contractions and, as it passes through the anus, is able to produce powerful stimulation of the mucous membrane. . . . The retention of the faecal mass, which is thus carried out intentionally by the child to begin with, in order to serve, as it were, as a masturbatory stimulus . . . is also one of the roots of the constipation which is so common among neuropaths" (SE 7:186–7).

For over two thousand years, metaphysics has been holding its stool back, accumulating its fantasies of immutable being. Now, it is finally having a dump, powerfully stimulating its rear end (*the end of metaphysics* is its rear end and the corresponding anal fixation stage; the hypothesis that metaphysics has ever reached the more mature phallic or genital stages, presupposed by phallogocentrism, is questionable) and experiencing pleasure in world-destruction. Mind you, having or taking a dump is not giving, or, if it is, then giving in the mode of withholding. The gift of indigestible stuff that comes out of the metaphysical anus perverts the logic of the excremental gift. Passing the stool, metaphysics does not let the prolonged moment of its "relief" pass. Does it want to have its pleasure nonstop? Does it wish to keep its masturbatory stimulus forever in play?

The desire of metaphysics in its most recent technicist permutation is still that of the premodern variety. Fecal mass is the model of mass, of matter as mass to be excreted in its entirety by spirit. Metaphysics contradicts itself and thrives on this self-contradiction: it wants release and retention, the never-ending pleasure of having a dump with the great fecal mass the world is or has become *and* total purification, the end of matter's passage whence it procures anal stimulation. The endless end of metaphysics captures the contradiction in its ideal state.

Moving on from the gift-image to the baby-image of excrement: whether the timeline of the Anthropocene begins at the dawn of agriculture or with the invention of the steam engine, the geodump and the aerodump it has created are the monstrous babies of progress. Just as the contents of a child's bowels "are clearly treated as a part of the infant's own body" (SE 7:186), so the dumped residues of human history are clearly treated as the issue of our technobodies extended in space and in time. In parallel, the industrial model of human reproduction, the production of babies on a national or global scale in the service of an idea or capital, is the augmentation of human biomass. Freud's children-as-sexual-researchers treated excrements as babies that have separated and fallen from their bodies; our age relates to babies as to excrement, dumped into the world. An infantile theory of what babies are and where they come from has sadly received historical validation. Augustine's *inter urinas et faeces nascimur* ("between urine and faeces we are born"), which Freud cites as a piece of evidence for the saint's inability to rid himself of obsessions with sexual life (SE 7:31), is in need of emendation: *qua urinas et faeces nascimur* ("as urine and faeces we are born").

Repugnance felt at the sight and at the thought of the Anthropocene is not a direct negation of anal metaphysical pleasure derived from world-destruction. The unconscious, after all, is unacquainted with *no*. This repugnance is a displaced and overdetermined affective reminder of excitation: "feelings [of disgust] seem originally to be a reaction to the smell (and afterwards also to the sight) of the excretory functions; and this applies especially to the male member, for that organ performs the function of micturition as well as the sexual function" (SE 7:31). Revulsion from the Anthropocene and, by implication, a misanthropic view of the culprit, *anthropos*, are the obverse of metaphysical pleasure, the obverse that stalks metaphysics and is thronged with bitter regrets, for it cannot have that pleasure for itself. We come up against accumulated detritus in the strata of the earth in a resting position of the erection lost, desire ostensibly pacified. Nevertheless, this layer cannot help but remind us of the seemingly interminable passage of "the column of faeces [that] stimulates the erotogenic mucous membrane of the bowel" and "behaves just as a penis does to the vaginal mucous membrane," the erection intact (SE 17:84). That which has colossally fallen walks us back to the verticality of the fall. Disgust at the deteriorating state of the ecology barely hides becoming-wet with pleasure. The Anthropocene complex is an affective disturbance of unconsciously wishing to partake in the tabooed excitement of world-destruction.

The meaning of the fecal gift is sacrificial, an echo of the burning transformation of the material world and life itself into biomass: "Faeces are the child's first *gift*, the first sacrifice on behalf of his affection, a portion of his own body, which he is ready to part with, but only for the sake of someone he loves" (SE 17:81). The subject of metaphysics is not prepared for self-sacrifice (in other words, for the sacrifice of its finite self), and so it preemptively declares that it has no body, that corporeality can fall away from it at any moment, as useless as a pile of excrement. It sacrifices its material existence in thought, so as to keep itself going in deed. But this is only a defense mechanism betraying a deeper (neurotic) fear of losing oneself, that is, of finding oneself at the final Shakespearean stage "*sans* everything."

Unconscious dread of the kind is another bond, tying the subject of metaphysics to those disgusted by the Anthropocene. The metaphysical solution of choice here is giving without giving, releasing a great mass of excrement while *taking* a dump with the world on the world. Or, as English idiom has it, shitting without giving a shit. Transcendental indifference and the theatrical gestures of letting oneself go before recovering oneself at home after a protracted odyssey are linked to the logic of that ingenious solution. Reclaiming, at the symbolic level, the residua of our economic activities in the Anthropocene, we join in the act, if only under the sign of repression.

A take-home lesson of psychoanalysis is that, shielded from interpretation, unconscious solutions exact a high price, complicating the original condition and creating new impasses. Efforts at tackling environmental problems and clearing up a technologically produced mess by means of technology are a case in point here. Geoengineering is a symptom of the geodump, not a cure. Attaining the reverse of the consciously intended, its *less* is *more*: it makes the dump grow, in the process of trying to unclutter the mess. Freud's proposal? "The talking cure": let's talk about it: let's put *logos* back into our eschato*logical* and scato*logical* dread!

falling in love and being dumped

Since we have already touched on the subject with Freud, we might as well contemplate the dynamics of falling in love and a peculiar falling out of it whereby I am dumped, or I dump, the other. I may fall in love, but, before consciously registering this event and before the burgeoning of consciousness itself (before the emergence of this *I*, therefore), love observed under the psychoanalytic microscope is a condition in which I let something fall from myself. It sacrifices a part I believe to belong to my body for the other's sake, causing that part to fall in the other's direction and charting with it a path the rest of me will follow. My falling *in* love reprises the fall of a chunk of me, a token of my being, *toward* and *for the sake of* the loved one. The ironic bit is that the first loving sacrifice hands the dirtiest and most superfluous contents of my body to the other. The child's *hoc est corpus meum* is excrement. Does this ever really change once the child matures? How can I know that the time, energy, and life I sacrifice for the other are not given and not received in the same manner as that initial gift?[1] Especially because the excremental present predates conscious knowing.

Falling in love re-embarks each time on a journey of the immemorial first offering: one falls away from oneself toward the other, not into the other. Tumbling from myself into love does not steal me from myself; on the contrary, it dispenses me to myself in the no-man's, no-woman's land between me and the lover. This inbetween is a relation that singularizes me outside the usual stricture of personhood and individuality, neither by imprisoning me in myself nor by capturing me along with the other in a claustrophobia-inducing dyad. I receive a form, which, nowhere near static, is only appreciable insofar as I continue to fall away from myself. If the feeling is mutual and the other is also falling toward me, then we are both suspended between two more or less fictional identities without intersecting at the same midpoint. We share the fall itself, plunging not hand-in-hand, shoulder-to-shoulder, but opposite one another and finding ourselves in a third element. In love.

Dumping a lover scrapes the symbolic coating off of the first gift. Brutally honest, the act calls things by their names, or, perhaps, calls things without resorting to naming, by pointing out with an accusative finger, categorically, the excremental sacrifice as what it is and what it has always been behind an array of sublimations. (On an expanded scale, this is what a genuine post-ideology looks like.) The dumped ex-lover is reduced to a pile of shit, which lay at the unconscious origin of the sacrificial offering. Dumping is the traumatic truth of love. The unconscious has taken the gift to be the giver's actual body part given for the sake of the other. Brutal rejection consciously identifies the whole body and the very existence of the giver with excremental mass. Dumping the other, I make that which falls toward me fall away from me. I also dump myself, seeing that both my form and that of the dumped ex-lover stemmed from our falling in love. We lapse to amorphous waste, a trace of past psychic metabolism restituted to, or reconstituted in, excrement.

To fall in love with and to dump the earth: is that possible? Freud would diagnose in the love one professes for the planet an iteration of the oceanic feeling, the affective identification with a diffuse, not yet individuated other. For the psychoanalyst, it is as impossible to love the earth as to dump it, because oceanic affect has not yet reached the stage of anal symbolism. Assuming that one can really fall in love with the planet, however, the situation is bound to be asymmetrical. While I, or a representative part of me, fall toward the earth in a loving sacrifice, the destination of my fall remains that toward which and onto which things fall, physically and metaphysically. The "gifts" of the earth (say, the crops it yields) may, to be sure, appear as the reciprocations of my fall toward it. And, in any case, reciprocity is not a necessary condition for, only an occasional supplement to, falling in love. So, the slogan *Earth: Love It or Leave It* is meaningful in the precise sense we are exploring here. It should only be added that leaving it (hence, leaving that toward which everything falls) is, in effect, archi- or meta-dumping.

Falling in love singularized the one who fell in it; dumping a romantic partner disfigures both parties to a relation. Dumping drops the other in a lump of useless mass. Its total rejection gravitates toward Julia Kristeva's abjection, whereby the subject "rejects and throws up everything given to him—all gifts, all objects." Having rejected the world, "a blank subject [*sujet nul*] . . . would remain, discomfited, at the dump for non-objects [*dépotoir des non-objets*] that are always forfeited, from which, on the contrary, fortified by abjection, he tries to extricate himself."[2] An ex-lover is obviously not equal to the whole world, but the unconscious might not see it this way. Dumping him or her, instead of merely terminating a relation, one acts on the unstated premise that the equation is valid. The massification and instrumental depreciation of a previously significant, cathected other are indicative of a problem in the depreciator her- or himself. Kristeva's blank subject is a psychic desert and a dump "forfeited by abjection" after it has forfeited every object. The dumped ex-lover is thrown out and the dumper is dumped by the act of dumping that refuses everything, cutting, besides this or that relation, the ties of relationality.

We would make a faux pas we were to interpret the abrupt transition from falling in love to dumping an ex-lover as a pendulum swinging from individuation to deindividuation and back to individuation in the next romantic involvement. *For one*, both are plunges or descents. Both tend away from oneself (away from one self) and revolve around "the same shit": the excremental offering to be accepted or rejected. In the case of falling in love, I admittedly acquire a dynamic form, whereas being dumped, I am reduced to an amorphous pile of detritus. Yet, the singularity that originates in love is on the hither side of individuality. Falling in love and dumping a lover bypass the individual altogether. *For another*, the undoing of relationality is irreversible. In another case, with another affair, there is no mending the threads that were torn when an ex-lover was massively dropped, released to the dump of being and the world, from which falling in love, the fall that love is, had temporarily sheltered parties to the relation. A subsequent fall will not fill the lack this event has left in its wake. A new infatuation will transpire within a negative space, the hole punctured by the act of dumping.

This thesis turns out to be especially consequential as soon as we ask what one falls in love with and what one dumps. In *The Meaning of Love*, Vladimir Solovyev reaches a surprising but cogent conclusion as he analyses the nature of fetishism. It is well known that fetishists are excited by a particular organ, piece of clothing, or accessory, putting a psychically salient part in the place of the whole. "But," Solovyev quips, "if the hair or legs that excite a fetishist are parts of a female body, then this very body in its integrity is but a part of female being. . . . So, what is the difference here? Does it matter that an arm or a leg cover a smaller surface than the entire body?"[3]

What one falls in love with is a fetish—a body part substituting for the body, a body replacing the sexed subject, a character trait standing in for the person, and so forth. Broaden as you may the circle of conspicuous parts, you will still fall toward a fetish, a singularity that stands out from the hazy background of the world as a dump. Already as infants, we send a protofetishistic part of ourselves to the loved other in a fecal self-sacrifice. And, if we are to believe the psychanalytic claim that sexuality is human ontology *tout court*, then our consciousness is no less fetishist than the unconscious, drawn to parts jutting out from our surroundings and claiming our attention. A humanly construed reality is a fetish of reality. Defetishizing it, we do not unmask the truth but descend into a dump, irreal and dehumanized.

In the sciences, an "objective" approach to "the object in general," *l'objet en général*,[4] does the trick of defetishization. In intimate relations, dumping an ex-lover instantaneously transforms him or her into a mass that merges with the fuzzy backdrop of overall insignificance. The scientific and interpersonal construction of objectivity becomes possible inasmuch as the subject is dumped. The caveat is that defetishization is carried out by means of a fetish, a part that stands out from the whole: numeric measures that give access to the scientific object in general and the thing at the root of all loving self-sacrifices, with which the dumped one is associated, namely excrement. Defetishization is disavowed fetishism. Anyway, it is enough for this process to have run its course once in order to open the scientific field of generality and to disband relationality in shit. Such "dump horizons" for experience predominate over and outweigh anything that, and anyone who, appears on them as already disappeared, done for.

When the meaning of being is being dumped, tinges of disaffection are evident at the ontological level, as well. The love affair with *to be*, the affair that is existence itself, is over. The intimacy of beings and being, the infinite fall of finite beings into the being that they are and that singularizes them, has been interrupted. Nothing severs *conatus essendi*, the Spinozan striving to persevere in being (E3P6), more effectively than the indifference and undifferentiation of the dump. The radical equality of dumped beings is that of nonbeing and death.

on the arcane utility of the useless

Dumping a romantic partner, one behaves toward a human being worse than toward a threadbare object: stripped of value and worth, a candidate for dumping is mixed with garbage. Indeed, while advanced capitalism assigns monetary figures even to things and activities outside the economic sphere per se, hyperdevaluation seems to be a constant of the dump across its many incarnations. The refuse is refused, thrown out and away, rejected. But the drainage of value from the dumped is not without utility. On the one hand, the logic of devaluation furtively orchestrates the unscrupulous manipulation of the devalued and the use of that which is deemed valuable. The wasting throw derives energy from the fall of what it massively drops. On the other hand, the state "without value" may be a frozen, indefinitely suspended stage of transvaluation, the uncertain and unfinished transition from a past network of stale values to a supple valuation to-come. Be this as it may, another positivity hides under the cover of negation.

The mass that is thrown away is not just retrieved after its fall and reintegrated into the circuits of utility, as in recycling. It is used *insofar as* it is thrown out, wasted, rendered useless. Market economies openly admit to these practices. Cyclical, state-trading, market-expansion, strategic, and predatory-pricing dumping "cover the practice of exporting at prices below the cost of production."[1] Purposefully squandering a part of the value embedded in the commodity, they sacrifice something they must write off as a loss for the sake of future possibilities. Devaluation begets a surplus invisible in the circle of capital C-M-C', its time and order irreducible to that of production and consumption (including productive consumption that gradually depletes the value of the consumed commodity in the production process itself). Its benefits, unquantifiable on regular balance sheets, need to be seen from an outside perspective of *quasi-transcendental accounting*, underwriting the economic system as a whole.

What presents itself as pure loss in dumping is a speculative gain of future market shares and monopoly powers. In this sense, Georges Bataille's conclusion that "the extension of economic growth itself requires the overturning of economic principles"[2] is perspicuous. It is, however, foolhardy to think as he does that "if a part of wealth (subject to a rough estimate) is doomed to destruction or at least to unproductive use without any possible profit, it is logical, even *inescapable*, to surrender commodities without return [*céder des marchandises sans contrepartie*]."[3] "Without any possible profit" and "without return" is how dumping would appear on the balance sheets of Bataille's restricted economy. Surreptitious expectations of profits and far-reaching returns beyond a given investment, production, and consumption cycles motivate this practice. General economy's lavish expenditure and pointless luxury overflow the "margin of profitless operations,"[4] reclaiming usefulness behind our backs through a good-old cunning of reason. They are governed by the rules of an accounting game that converts losses into deferred gains, sheer playfulness into hard work, expenditure into investment, the minuses into the pluses.

The apparent waste of economic dumping is comprehensible within the logic of capital bent on conquering a foreign market. As prices plummet, a mass of devalued commodities is dropped onto the consumers in a bid to establish a purchasing pattern and dependence on the item in question. Such machinations are then supposed to give the producer a future competitive edge in marketing and retail. What is the philosophical meaning of selling below production value? Doesn't it, in the short run, affirm the non-reproducibility of capital, an economic death of sorts? Or, in a patently Hegelian turn of events, does it not draw strength from death, building upon the negative power of finitude and occupying the position of the master, expressed in subsequent market monopolization?

Dumping puts general economy at the behest of restricted economy and extracts surplus value from devaluation, a buoyant afterlife from the demise of immediate vitality. Not to stay behind, "antidumping legislation" contrives a paradox of its own. In the United States, after the Department of Commerce (DOC) receives an affirmative preliminary finding on dumping from the International Trade Commission (ITC), it issues questionnaires "to 'mandatory respondents' — the largest known foreign producers and exporters of subject merchandise from the countries in question." These are so extensive that "responding to an antidumping questionnaire usually requires the diversion of significant company resources and retaining legal, accounting, and economic expertise."[5] DOC fights fire with fire: sorting out a dump, it multiplies that which it opposes, much like geoengineering that contributes to the growth of the geodump. Were it earnestly to tackle dumping, it would have had to scrutinize and subvert the theoretical and practical grounds for capitalism.

To translate Marx into the language of neoliberal trade law: under capital's regime, labor is *always* a dumped commodity, no matter the "fairness" of trading. So long as capital exists, wage laborers permanently bear the brunt of devaluation and sell their labor power below its full value. (With the rise of the precariat, they do not even receive enough for the reproduction of the conditions of production, getting back less than "the value of the commodities which have to be supplied every day to the bearer of labour-power, so that he can renew his life-process."[6]) The existence of capital implies ipso facto that workers are dumped. Surplus value, the portion of labor's value that instead of returning to the worker is rerouted toward capitalist growth, is equivalent to the dumping margin, "which is calculated by subtracting the export price from normal value and dividing the difference (assuming it is positive) by the export price."[7] Try plugging wages into the spot occupied by *export price* in the equation, and the total value created by labor in place of *normal value*. You will obtain the real dumping margin of capitalism, for which the one and only antidumping measure is communism, an articulation of the common that resists the disarticulation of the randomly piled up.

In the commercial practice of dumping, businesses temporarily put themselves in the shoes of their employees. They throw a chunk of themselves away so as to recover more than the capital that has been sacrificed on the scales of economic power and potentiality. The workers, for their part, do not see their time and labor boomerang back to them as a company does. The value that, diverted from them, forges and reinforces the means of their oppression falls away as an undifferentiated and indifferent *mass*, a word Marx frequently avails himself of in *Capital*. The growing massiveness of capital is a dump wherein the workers are dumped, time and again, irrespective of their unique skills and the actual labor time they have put in. But there is another dump, its unstable contours dogging the fringes of the first. That is the industrial (and now the postindustrial) reserve army.

Comprised of the unemployed and chronically underemployed human masses, the reserve army of workers fluctuates in tandem with capital's productive needs, swelling or diminishing according to the phases of economic expansion and contraction. *Within* the economic process, excess labor force is useless; from the *relative outside* of capitalist economy, it plays a vital role: "The industrial reserve army, during the periods of stagnation and average prosperity, weighs down the active army of workers; during the periods of over-production and feverish activity, it puts a curb on their pretensions."[8] The mass of those eager to work yet denied employment prompts wages to fall in inverse proportion to its growth. This mass weighs active workers down, as Marx has it, further dumping the wage dump. So useful is the increase of human disposability, dispensability, and formal uselessness for capital that it "can by no means content itself with the quantity of disposable labour-power which the natural increase of population yields. It requires for its unrestricted activity an industrial reserve army which is independent of these natural limits."[9]

Hard and fast divisions between an active and a reserve army of workers—the inner and the outer dumps—have become fainter over the century and a half that has elapsed since the publication of Marx's magnum opus. The relative outside has crushed into economic interiority: deprived of job security and forced to work on a short-term contract basis, to offer volunteering or unpaid training services, and to toil for perceived benefits other than a wage, flexible labor force is a reserve for capital to dip into as it sees fit. The reserve's mass weighs upon the "active army of workers" from within. Labor suffers continuing devaluation, licensing capital to derive *its* value from the unrestricted use of the formally useless. Falling lower still, the dumped commodity *par excellence* is subject to surplus-devaluation after having been utilized. The extraction of value from the devalued, from the energy released upon impact in being dumped, is by now the template for instrumentality and use in general.

Squalor and the dreary living conditions that dredge up another sense of the dump are part and parcel of commodified labor's massive throw. In the eighteenth and nineteenth centuries, proletarianization occasioned urbanization, a rapid relocation of agrarian populations into the new overcrowded centers of industrial production in cities that largely resembled shantytowns. The masses of industrial workers dumped in urban areas experience "such negation of all delicacy, such unclean confusion of bodies and bodily functions, such exposure of animal and sexual nakedness, as is rather bestial than human," in the words of an 1866 report on the state of public health Marx quotes.[10] Applied to the fledgling proletariat, discursive bestialization is a familiar leitmotif in classical political economy, notably in Joseph Townsend's 1786 *Dissertation on the Poor Laws*.[11] And, in fact, the report that has made its way into Marx's *Capital* tacitly frames the propensity to bestialize the working class in the obscenity of the dump: humanity reduced to its bodily functions, a "confusion of bodies," "such negation of all delicacy."

Postindustrial gentrification of core city areas relocates the dump elsewhere, for instance to the suburbs—as in some of the Parisian *banlieues*—and globally redistributes it to bustling manufacturing hubs, such as Bangladesh. As mobile as capital itself, the urban and suburban dump is both horizontal, often coercing strangers to cohabit in the same apartment, and vertical, piling up families and individuals on top of one another in high-rise buildings. In the worst of cases, it throws people out into the streets, homeless and exposed to the elements. In all others, it furnishes for bodies the kind of housing that is not a dwelling.[12]

Little wonder that the concept of use has come to be so contorted. We lack direct access to utility for "genetic" reasons, because a thing is necessarily useful for something other than itself, necessarily pointing beyond its body in being used, and for historically contingent reasons, because nowadays quantitative considerations and relations of exchange mediate usability. Viewed through the camera obscura of capital, use has no use in itself, and apparently useless phenomena are crucial to value augmentation.

Singular purposes and ends, that for which something is, disappear from the dump. If we are to have recourse to Nietzsche's language, the dump is beyond good and bad, as well as beyond good and evil, in that the practically inflected question of the good—what is this good for? whom does it serve? *cui bono?*—does not arise there. Nonetheless, the act of dumping has a metapragmatic use, be it as minimal as obscuring these inconvenient questions and the notion of the good itself in the universe of capital. In short, the technocratic, efficiency-driven, economicist attitude posits uses without the good, a concept repulsed for its allegedly teleological, obsolete connotations. It blots out a basic intuition that use without the good, which would be infinitely ramified in its desired purposes and realizations, is good for nothing.

the portrait of a thing as its own wastebasket

What good is it to multiply instantly discardable materials for wrapping or carrying consumer goods?[1] As visitors to the United States will observe, purchases are double-packaged there with almost obsessive zeal. From paper coffee cups to plastic grocery bags, you leave cafés and supermarkets with a not-so-symbolic excess in your hands. You may cite pragmatic reasons for this practice. Doubling paper cups keeps your fingers from getting burned by a hot beverage (though an extra cardboard sleeve—a part-object that with its own incompleteness highlights the deficiency of the first—may be marshaled for this purpose). Putting one bag inside another bolsters it so that it would not tear open as you carry your haul home, and so forth. But even such simple explanations betray something other than matters of convenience.

In a culture of obsolescence, of goods already produced and consumed as garbage, things no longer work as they should. To fulfill their functions adequately, they must rely on supports and reinforcements provided by other things, often of the same kind. Taken in isolation, a paper cup fails to hold the liquid it contains and a plastic bag disintegrates, which is why an equally shoddy product is added to boost them. Strength in numbers compensates for individual frailty: neoliberal capital transfers onto the arena of objective commodities the same principle it denies the workers in their professional lives, by pitting them against one another and by interfering with the unionization of labor.

I am aware of the silent *complicity*—literally: a folding-with and, thus, a kind of behavioral double packaging—between the industrial generation of consumable trash and the consumers' own refusal to commit to reusable objects that would cut down the wastefulness of our lifestyles. (This, too, is gradually changing, at least when it comes to shopping habits; in the fall of 2014, California became the first state in the United States to pass a law banning single-use plastic bags.) That said, what interests me is how the phenomenon of double packaging, taken for granted in the American consumer universe and in its metastatic extensions to other parts of the world, encapsulates the dump above and beyond the antagonism between economy and ecology. How do we navigate the bewildering doubling of the double and the single, an occurrence or a thing repeated in itself and a disruption of the repeatable horizon for experience, clogged with garbage, due to this material redoubling? How do double bags and cups create a single-use world where finitude means being finished, *è finita la commedia*?

Before it rips down the middle, seize the ends of an overextended cord, stretched between the single and the double, one in two and two in one. You will then lay hold of something like this:

1. *Macrosingularity.* Single-use bags, cups, tissues containing plastic fibers, and worlds send us back to Heraclitus's fragment 89: "Those who are awake have one common world-order [*koinos kosmos*], but each of those who are asleep retires to his own, private world-order [*idios kosmos*]." This fragment resonates, as an uncanny double, with the one we've already stumbled across, "Just as a heap of refuse [*sarma*] piled up without purpose, so [is] the most beautiful world-order [*kallistos kosmos*]." A dreamworld is isolated and isolating. In the dark of sleep, it shines on each dreamer separately, for the benefit of each alone, or not even.[2] Were there nothing but fragmentation, the world would have been a pile of refuse. (*Idios kosmos*)∞ = *sarma.* Single-use double bags and double cups are gateways to the dreamworlds of private consumption, and they superimpose those worlds onto the common, onto ecology, which they fracture and imprison in the idiotic (*idios*) isolation of the heap. At the macrolevel, the dump consists of fortuitously accumulated singularities alienated from one another and from the cycles of planetary metabolism. Hence, the temporal asymmetry between the wasted-yet-enduring

artifacts and environmental processes: a single instant of use is followed, in the case of plastic, by centuries of decay.

2. *Microduplication.* To schematize a little, once you peel off one layer of packaging, you will discover, in the style of a Russian doll, a second identical package in it. A double bag (doesn't the expression itself meld two into one?) and a double cup are a thing dumped within a thing and presumably holding the "real" thing, be it groceries or a hot beverage. Their contents are thrown within that which is thrown into its double in anticipation of being thrown out as soon as I finish drinking or bring groceries home. The self-replicating dump is a shell to be discarded when I retrieve a valuable kernel from it. But not so fast! The kernel is not enclosed within a shell; as I approach it from the outside, I find an empty shell within a good-for-nothing shell, a desert growing by accretion around the goods I have acquired.

A double cup or a double bag is a thing as its own wastebasket, dumped into itself. The dumping throw is massive: commencing with one piece of packaging, it increases by a factor of two and is then multiplied by an *x* number of identical nonreusable items. The contents thrown or poured into this volatile, rapidly spreading, self-duplicating *one* are poured or thrown into a thrown thrownness. Superficial casings tell us more about ontology than the precious contents they hold. It is likewise with bottled water swarming with microplastics, most of them derived from the plastic cap and the bottle themselves. The container is dumped into what it contains; we indeed drink "the bottle with the water."[3]

If frenetic piling on is already happening within the one bag or cup that nondialectically divides into two (I am, of course, alluding to Mao Zedong's principle *yi fen wei er,* "one divides into two"), then its social, environmental, and global consequences are so much more disastrous! Note that the process of division is not just reproductive. Nor is it an upshot of iterability, repeatability, the possibility of reinscribing or remarking a singular mark or object. Riven sameness does not leave its own side in its otherness to itself. The doubling of a container thrown into its identical twin—therefore, virtually into itself—before other objects fall into it is akin to atomic fission, which also proliferates, becoming eerily generative. Is this an external, material projection of my self-relation, of *I* and *myself* who is the same as, yet also other to me, each too frail to uphold an identity?

On the surface of it, double bagging and double cupping epitomize the material interconnectedness of things, whether we sort them under the heading of *res extensa* or *res cogitans*. Silently, wordlessly, doubling conveys that nothing and nobody exists in isolation from others, that objective and subjective autonomy is illusory. Nonetheless, in its consumerist single-use rendering, it conjoins two objects that are the spitting images of each other, a cipher for the homogenization of our world. The same combined with the same is, precisely, the model of the real, of (mass) communication, and of thinking, foisted upon us in the age of the global dump. Physically articulated, the two are ontologically disarticulated, heaped up in the climate of obscene obsolescence, dumped. Their doubling is the opposite of Luce Irigaray's tenet "to be two," which implies sharing in and across difference.[4]

Should there be any lingering doubts that we are dealing with more than just paper and plastic, we would do well to recall that doubling is the sine qua non of signification. The broken unity of the sign consists of the signified thing and the signifier, standing for, replacing, and in the last instance substituting for the signified. In Émile Durkheim's *Elementary Forms of Religious Life*, a precursor for the splitting of the sign is the mind-set of ancient animism, where a thing is both itself and a receptacle for *mana*, the spirit that gives it life and makes it what it is.[5] A site of *mana*, the tree is both a tree and not a tree; it is itself and an excess over itself. This also applies to human meaning that, overflowing the materiality of the body it signifies, overwrites the self-significations of matter, to which Derrida affixed the neologism *archi-writing*. The world is enchanted because it is brimming with meaning, welling over with significance—in other words, because beings within it do not fully coincide with the rigid images (ideas, concepts, etc.) of themselves.

At the height of disenchantment, the doubled is replaced by itself, by its exact replica, despite being physically present, in attendance at the site of the supplanting. As soon as the excess of material doubling is incorporated into the thing, the latter loses its meaning. One thing is deficient, useless, and meaningless. More so still is the one thing added to exactly the same (thing).[6] Now, since the double bag syndrome touches human life at its core, what it says about our reality is that we, perceived by what Marx calls "the automatic subjectivity" of Capital as walking-taking bags and cups, are replaceable in relation to others, who are presumably our doubles. The promotion of teamwork and cooperation at a capitalist workplace does not valorize mutual interdependence and the sharing of differences; it fosters prosthetic supports of some parts in defective human capital by others treated as exactly the same in the shadow of redundancy.

Holding a cup within a cup in your hand, you are expected to throw this hardly noticeable double thing away when you are through with your coffee. You will act well if you recycle it (or them). But you will have been preempted. Prior to any action on your part, you receive from a helpful *barista* a thing thrown away into itself and subtly inviting you to throw away your future, plus that of the liveable planet. The cup becomes its own dump, its own garbage bin, disposed of together with the remnants of meaning it once had. Its diminutive size notwithstanding, it is a wastebasket for the person crafting your beverage, for you who drinks it, and for our shared environment. More than a hot beverage, that is what you *consume*.

Dumping is nowadays the most lucrative activity of all. Within the capitalist logistics elucidated by John Maynard Keynes, wouldn't it be more profitable to triple or quadruple things, inserting a cup into a cup into a cup and letting the desert-dump grow exponentially? Continuing on this path, we will transform our planet into a dumping ground at lightning speed and throw ourselves away more efficiently. Then, again, there are always unpleasant trade-offs for economic growth and job creation!

It is not enough to criticize, with righteous contempt and derision, the wastefulness of double packaging and its adverse effects on the environment. Far from an aberration, the phenomenon exposes essential elements in the makeup of things and in the human condition at the current historical conjuncture. Things are not self-sufficient and neither are humans, whose existence, to boot, has the character of being thrown into the world for nothing, potentially wasted. The question is how to cope with this endemic interdependence. We cannot put an end to the doubling of signification any more than we can predict the exact curve of the toss that is a human life. But we can hold out against the tendency to duplicate the body of the thing at the price of its meaning and environmental viability. With all our might concentrated in the biomassed power of not-nothing, we must press against the fatalistic determination of the existential throw, steered toward the finality of being thrown out, used up, done away with.

dumpology

There will never be a "Newton of a blade of grass," Kant wrote in *Critique of Judgment*.[1] Will there ever be a Linnaeus of the dump?

Kant reached his verdict on the assumption that natural generation is incomprehensible if one's thinking relies on the laws of mechanical causation alone. Just as the principles of physics do not fully explain biological processes, so the rules of ontology are inapplicable to the unbeing that we inhabit and that inhabits us. A Linnaeus of the dump is an oxymoron. In virtue of what it is, in virtue of evading being, the dump is not available for discernment and classification. Leaving no place for place, it disallows taxonomies and typologies where discrete categories of leftovers would find their niches and where they could be described, studied, cataloged, deposited, or discarded. It is incompatible with *logos* in its many manifestations: dump-study, dump-speech, dump-voice, dump-articulation, dump-discourse are nonexistent. A surrogate for ontology at present, dumpology is dump~~ology~~.

It is impossible to name, study, and classify the dump's contents. Yet, we do nothing but name, study, and classify them. There is nothing else to do, nowhere else to turn or to return anymore, and so each of us is an amateur Linnaeus of the impossible. In it, in the dump, we scavenge for knowledge and for things, for breathable air, drinkable water and something edible (including "food for thought"), for experiences, for meaning, for a future. We pick up the scraps of these but do not pause to consider, in the few and all too brief instants of respite from our searches, the way they hang together in their insularity, interlaced by the dissociation of the global dump. In a certain manner, then, the lines of demarcation between the aims of our rummagings sketch the grid for a table of classifications, a taxonomy of the dump. Whatever we are doing *in* it, whatever we are imagining, desiring, assimilating, or recoiling from, we do *to* it, ephemerally shaping its amorphousness from within, absent the boundary between the inside and the outside. We are not trailblazing but trudging a dirt road or following a flight path of airborne debris, one of countless trajectories crisscrossing the dump and receding back into obscurity.

The word *dump* is, as we have seen and heard, a dump for multiple significations, a hodgepodge of heavy, falling, massified, vulgarized meanings indifferent to their differences. We have been fumbling for some semblance of order among anonymous and impersonal discernments and protocols, keeping in mind that this would have been the discernment of that which denies discernment. An attempt at classifying mental, corporeal, metaphysical, construction, chemical, nuclear, and other species of debris separates in thought what is mingled in deed. A foredoomed undertaking, if there ever was one, it veers perilously close to idealism. We are like someone who, while conscientiously sorting out refuse and puzzling how to fill preexisting boxes with it, invents new trash and recycling bins: pink, orange, purple, beige, in addition to the usual blue, yellow, green, and black.

The slightest differentiations undump being. They dare broach again the Platonic question—before its appropriation by Christianity—of what should be saved, rescued from the dump, salvaged from the oblivion of a massive fall *and* from the frozen memory of metabolism at a standstill. That which might be saved need not be exempt from the status of garbage, so as to be torn from the clutches of the dump;[2] it only needs to be recognized for what it is. Seen in this light, a preparatory conceptual errand we need to run is disentangling the dump from cognate words: rubble and rubbish, trash and garbage, waste and debris, detritus and remains.

In and of itself, the luxuriant synonymy, the surface of which we have just scratched, is noteworthy. Languages draw fine, at times hair-splitting, distinctions and offer a profusion of kindred words where these pertain to things that matter most. Inorganic chemical processes and living organisms all leave residues in their trail, but humans are perhaps the only beings capable of feeling responsible for the waste we generate. I would go so far as to claim that garbage is a humanizing factor, which, as an object of concern and responsibility, bestows on humans our humanity. Dumping is, quite the opposite, shirking that responsibility and, with it, the task of becoming-human.

However impossible it has seemed just now, a budding dumpological taxonomy consists at minimum of the following categories:

1. *Waste, by-products*: the economic underside of spending and investment (for instance, of energy). Waste expresses the inefficiency of investment; by-products are the accumulated objective traces of a life activity or inorganic chemical processes.

2. *Garbage*: residue that is confused, garbled, and, therefore, impervious to sorting. It also connotes unusability and worthlessness.

3. *Dust*: desiccated garbage (as in the old-fashioned *dustbin*), blending inorganic matter with mold spores, fungi, pollen, and dust-mites.

4. *Refuse*: garbage that is unwanted, rejected, refused.

5. *Rubbish* and *rubble*: organic and inorganic wreckages, ruins; matter shorn of sense, dismantled into an absurd heap, worldless.

6. *Trash*: matter itself, taken in the deepest sense Western metaphysics assigns to it. If matter is originally vegetal—*hulē, materia, madeira, madera*—then trash is vegetal remnants, the fallen leaves and twigs, to which the Old Norse *tros* alludes at its root.

7. *Debris*: garbage as a collection of fragmented remains (for emphasis, the word is in a double negative: *de+bris*, breaking apart).

8. *Litter* and *flotsam*: the scatter of debris, some on land, the rest of it carried by water, floating.

9. *Detritus*: hints at the process that has culminated in debris: erosion, wear-and-tear.

10. *Dump*: a massive fall of stuff unloaded with unalloyed indifference, snowballing, swallowing all of the above into itself.

These are some of the types that may appeal to our inner Linnaeus. The energy of the dump and of its cognates flares up in the disparities between their verbal and substantive dimensions, between what they do and what they are. The acts, imparting to the thrown heterogeneous destinies and destinations, dictate their own taxonomy. Some gather the shattered pieces; others refrain from fitting them together. Some accept, while others reject the rests. Some are charged with the memory and pain of loss; others are conducive to out-and-out forgetfulness. Still, the dumping and the dumped are our main process and product, energy in motion and at rest, released and contained.

The transversal axis of trashing activities is metabolism, whether psychic or elemental, physiological or geological. Does the formal or material identity of a disposed, scrapped, discarded entity stubbornly endure? Does it undergo a more or less rapid *trans*mutation feeding a nascent life? Is the memory of its former unity emblazoned in the never-ending present of a mental-environmental trauma? Does forgetting compost it into fertile soil for new growth?

In ontological toxicity, metabolism is halted. Needless to say, this indigestion affects knowledge, as well. Despite sharing some attributes with other disposal methods, dumping is a category apart, standing out from the taxonomy it participates in. It is none of the other nine and all of them. As it proscribes *logos*, the dump strikes at the heart of articulation, meaning-making, and knowledge production. The second strike is the welling up of indifference and undifferentiation that do away with meaningful discernments. The third strike is the massiveness of the dumped pile, into which knowledge falls. Three strikes, and the dump is out of the sphere of cognition. Or, more to the point, it has devoured cognition whole, together with the illusion of independence our thinking nurtures vis-à-vis the dump it deems too undignified an object of thought. There is no ontological toxicity without the epistemic kind. And vice versa.

estamira, esta mira, "this sight"

If, for some strange reason, you type "landfill Washington DC" in Google's search engine, one of the top hits is a Yellow Pages (YP) link *Best 8 Landfills in Washington, DC with Reviews*.[1] Advertising campaigns do not give up when it comes to the dump, which they hand over to commercial rankings. Touched by commodification, systems of classification have become not only classed but also, themselves, classified.[2] The "preferred" link on YP is to Waste Management, a "small business dumpster rental [that] allows you to quickly find the right size dumpster for your project, schedule delivery and pay, all in just a few clicks."[3] The success of your endeavor depends on the efficient allocation of your trash to "the right size dumpster"; Prince George Universal Recycling follows closely behind, albeit apparently unconcerned with such singular details in its catholic aspiration to recycle the world. The madness of classifications that assign numeric ranks to virtually everything bleeds at its edges into the derangement of the dump.

What sort of a gaze can we train and cultivate in response to these extremes? We have glimpsed the horrified regard of Benjamin's *Engel der Geschichte*, the regard that is as superhuman as it is subhuman. In *Estamira* (2004), Brazilian filmmaker Marcos Prado presents an alternative: an elderly woman, who is the film's eponymous heroine and who lives off the scraps she gathers at Jardim Gramacho, one of the largest landfills in the world situated (until a few years ago) in Rio de Janeiro.[4] Estamira's gaze is that of lucid madness and deranged sanity; as she puts it, "if I could not recognize my own disturbance I would not be Estamira."[5] Mirroring the unworld of the dump, her vision also reflects upon itself, in particular upon her name, which, divided into two words—*esta mira*—means *this sight*, *this aim*, *this target*. "I, Estamira," she says, "am the vision of everyone. No one can live without me. No one can live without *esta mira*. [*Eu, Estamira, sou a visão de cada um. Ninguém pode viver sem mim. Ninguém pode viver sem esta mira.*] I feel pride and sadness due to this." The broken mirror of her reflection corresponds to the wrecked reality of the dump. In its sheer singularity, *this* sight is simultaneously general, belonging to everyone. Sponge-like, it has imbibed the indifference of the dump and has clung onto *this*, the difference of embodiment, incarnation, perspective.

Through the lens of *Estamira*, we see what Estamira sees and what *esta mira* sees. Opening shots: on her way from the shack where she lives to Jardim Gramacho, Estamira traverses a desert landscape of rocks and dust. Main title: against the backdrop of a stormy sky, shredded plastic bags and vultures soar in a wind-driven chaos, the ones indistinguishable from the others. A semi-naked human cadaver found in the trash and unhurriedly mantled with ubiquitous black garbage bags. A long-time friend who sleeps amid the rubble in close proximity to the fires fed by methane and other gases. A new batch of refuse dumped at night, threading in and out of visibility, occasionally illuminated by the scavengers' roaming flashlights. ("This here is a slave's masquerade. . . . Slaves masqueraded as free people [*Isso aqui é um disfarce de escravo . . . escravo disfarçado de liberto*]," Estamira comments her fellows' nocturnal searches.) A load of canned food carefully picked out from underneath mountains of garbage and saved for a much-vaunted pasta sauce.

I can clearly hear your murmurings: But these are not things I see day in, day out, if ever. Can Estamira's vision be mine, then? Moreover, her vision filters through the camera and the director, coming to me second- and third-hand. With so many superimpositions and distortions, where is *this* sight? Where are the points of access to its unique perspective with its profession of generality?

Estamira's vision would have been yours were you capable of seeing microplastics in the water and the soil; CO_2 in the air; pesticides, traces of heavy metals, and irradiation (cobalt, cesium, etc.) in foodstuffs. It will have been ours after environmental disaster zones have expanded from some pockets of contamination, from a few no-go areas, to the global rule. It is already everyone's underneath the thin veneer of consciousness, without us knowing it. The aesthetic prism, through which her reflection travels, is refracted, dispersing its dark light: this is not an unfortunate hindrance to be overcome. Only an inadequate, prismatic representation is adequate to a damaged, mutilated reality. Adorno's aesthetic theory testifies to this predicament of art that shakes off the temptation to reconcile in its insular medium a painful contradiction unresolved in the world. The lens is cracked, the frames do not align, and that is for the best in a situation going from bad to worse.

The territory of dump aesthetics extends beyond the represented content to the representing form, which is quite unremarkable, nothing special—a documentary where protagonists speak without any overt interferences by the filmmaker. The formal dump (itself a contradiction in terms) is in the bungled noncoincidence of aesthetic content and form, the *this* of Estamira's lucid madness and of her unworld, on the one hand, and the seemingly unaffected, unblinking gaze of the camera sending the viewers back to Dziga Vertov, on the other. Except that, already in the singularity of her sight, Estamira anticipates the generality of everyone's vision in its interface with technological neutrality. The jargon of technology is rife in her discourse: from natural and artificial remote control, through the "format" of the human, to the digital and electrical aspects of embodiment. She has imbibed the contradiction that will be replayed in the cinematic aesthetics of *a documentary on a landfill*: documenting the landfill and taking its nonplace on the landfill, cothrown with its subject.

Estamira is as much subaltern as she is "superaltern," nearly transcendental in epitomizing the conditions of possibility or impossibility for dump experience. In her singular generality, she is bordering on space, her filiation to neutrality steeped in sexual difference: "Estamira could be the daughter, the sister, or the wife of space. But she is not. [*Estamira podia ser irmã ou filha ou esposa de espaço, mas não é*]." Why not? Because, at once more and less than a relative, she inches toward how space would have seen itself and how it could have come to self-consciousness as a dump. "This sight" is the eyesight connected to the perspective of a certain human being—"I was born on the 7th of April 1941. Flesh-and-blood. The format. Format: human, even [*A carne e o sangue. O formato. Formato homem, par*]"— and to the unsettling scenes of devastated landscapes, airscapes, waterscapes, and even pyroscapes, referring to the permanent fires blazing in the dump. *Vista* and vistas, viewing and views, looks and outlooks, homonymous proper and common names confused, the one dumped into the other, Estamira's sight *is* the sight of devastation. The documentary medium enters the fray as the aesthetic consummation of these confusions, hovering between Estamira's sight and the sights of devastation. That is why the film can claim for itself, without appropriating it, the mission Estamira considers her own: "My mission . . . is to reveal the truth, only the truth [*revelar a verdade, somente a verdade*]. . . . Or, to teach and show what they do not know. The innocent ones. But there are no innocent ones any more [*Não tem mais inocente*]." There is no innocence because one sight is implicated in the other: a perspective in the death of perspectivalism and, therefore, of experience as such; the seeing sight in the seen; the dump of the sensorium in the dump of the unworld. There is also no innocence, because in the global Information Age there is no ignorance, only fluctuating thresholds of repression, of knowing enough in order not to want to know more. The dump is the epicenter of this complication, coimplication, and implicature of knowing with not-knowing.

Estamira's revelatory teaching proceeds by way of recollection through a series of strange cues to the goal of lifting the repression that safeguards the illusion of innocence. She understands her knowledge in a Platonic key as anamnesis, unforgetting: "Before I was born, I already knew everything. Before I was of flesh-and-blood. Of course, if I am the edge of the world. [*Antes de eu nascer eu já sabia de tudo. Antes de eu estar com carne e sangue. É claro, se eu sou a beira do mundo*]." Knowing is not registering new information, but receiving reminders of things one has known all along without consciously knowing them, a memory predating one's body and one's flesh. The impersonal character of this memory matches the rampant deindividuation of the dump.

Estamira's filiation to space looms again in the recollection of what has been never present and never represented: "Of course, if I am the edge of the world . . ." At the edge, forgetting and unforgetting are the penumbras of sight with its roughly subjective and objective connotations. They are the gateways for trafficking scraps of experience there where metabolism has been suspended. What exactly is it that passes through them, flouting the blockages of ontological toxicity? Estamira does not hesitate in her response: the vestiges of life and neglect, being and nothing. "This here is a dump of rests. Sometimes, it's just the

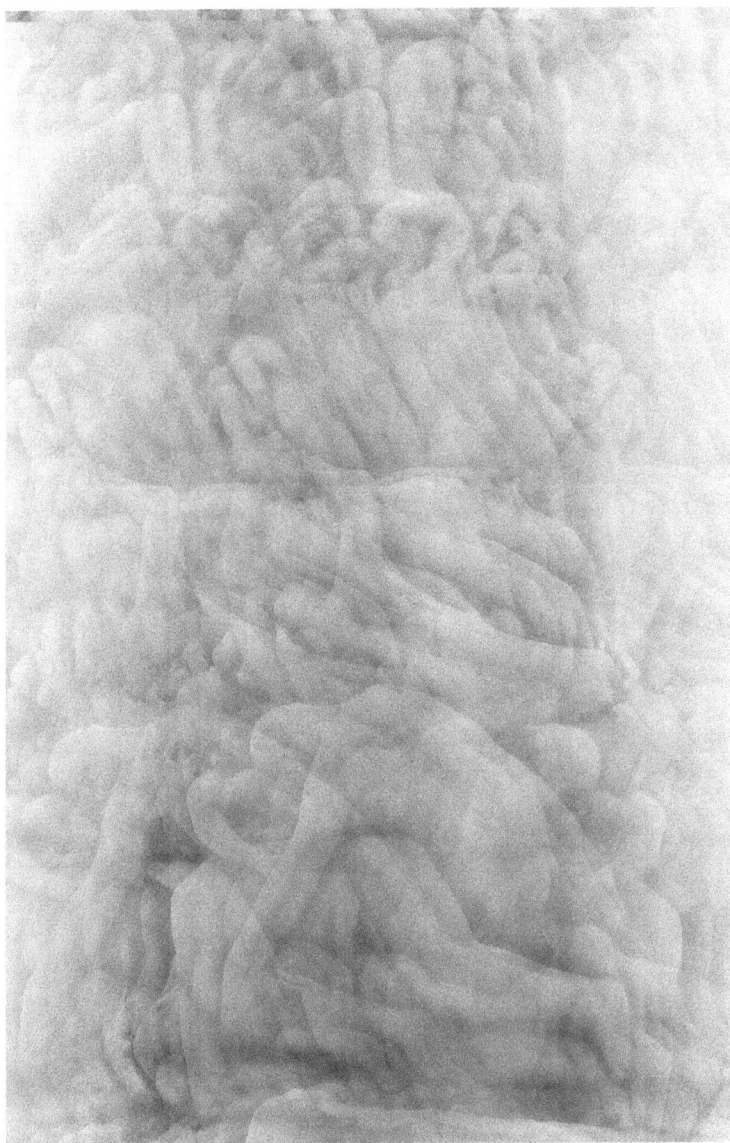

Figure 16 Drowning in, Anaïs Tondeur, 2018–20, Pigment print on Murakumo paper, 42 × 63 cm

rests. And sometimes the neglected also comes. Rests and neglect [*Isto aqui é um depósito dos restos. Às vezes é só restos. E às vezes vem também o descuido. Restos e descuido*]." Living in a dump is existing and subsisting on the neglected slivers of being. It is subsisting, without abiding, on neglect: being neglected and deriving one's being from neglect. In the rare flashes of lucid madness, which Estamira (and *Estamira* [2004], and *esta mira* . . .) embodies, living in a dump is shining light on forgetting and the forgotten, unforgetting without at the same time eclipsing the dark glow of carelessness and dereliction with the blazes of memory and representation.

Rather than face another world, the edge of the world in a dump slips into an unworld. There are no worlds without edges, blunt or sharp: the edge apportions to the world its worldhood.[6] Effacing the edges of things in their massive outpouring and fall, the dump eviscerates the world. Estamira is the edge that narrates the devastation of edges and, with them, of worlds.

the writing dump

Sacred writing is for us an indecipherable hieroglyph. It is inscrutable regardless of our insistence on composing fresh homilies and sermons (something that often indirectly happens in the name of moral or environmental philosophy), reverence and respect paid to certain texts as in Abrahamic religions, or interpretation of scripts from Sumerian, Mesoamerican, or Ancient Chinese civilizations. Sacred writing has vanished even from theocratic regimes, so long as holy texts coexist there with smartphones and the (often censored) Internet. The secular mode of inscription is the fused horizon both for those societies that embrace it and for those that vehemently reject, but are effectively ensnared by, it.

The roots of writing are steeped in sacrality. This is the case not only because its earliest object is divine veneration, but also, and more importantly, because its medium and the message alike are highly selective, set apart from the rest of everyday life, reserved for special occasions—setting down the law, to name one. Entrusted to a durable, costly, and labor-intensive substratum (stone tablets, treated animal skin, papyrus), sacred inscription is an exception from ordinary reality, its meaning also accessible to the select few who possess the skills necessary to read and interpret it.

As it undergoes secularization, writing capitulates to the power of literature. The literary paradigm insists on classifying any text whatsoever based on its genre, so that ancient epics, hymns, sutras, sermons, epistles, scriptures, and codices are all organized under the banner of religious *literature*. Within this paradigm, sacred writing is unfathomable unless we box it in the appropriate category according to the law of genre. But the act of slotting sacred writing into an overarching classificatory system reduces it to a part in a larger whole, negates its absoluteness, and dilutes its sacredness. Literature's preferred paper substratum is, moreover, easier to produce and to destroy. Less selective, literary writing encompasses a broad array of topics and, given rising literacy rates, is available to a growing public. Though sutured to the civil religion of nation- and identity-building, and though initially riven between "high" and "low" cultures, secular (literary) writing readily lends itself to democratization as regards its content, materiality, and circle of reception.

To complicate our otherwise plain-sailing story a little, we may transpose the deconstructive approach to speech and writing onto sacred and secular kinds of writing. What if, just as writing was *before* speech for Derrida, so secular writing is "first" with respect to the sacred? It would be easy to explain the priority of secular vis-à-vis nonsecular texts. As the negative form suggests, the nonsecular pledges a return to the sacred after the secular. To inscribe it in another shorthand, this detour charts the path of postsecularism. But that is not what I mean; my claim is, precisely, that the secular precedes the sacred, not the nonsecular. How so?

Mull over the word *secular*, which we have already located on Shakespeare's stage, for another brief moment. It speaks of the world, a place, a network of places, set apart in their habitualness from uninhabitable sacredness. It also speaks of time, of the generation, the age, *saeculum*, an epochal stretch of the here-and-now. Secular writing is worldly and temporal, the wrinkles of its age carved deep into its skin (or does its skin form and grow around the wrinkles?). It is the writing *of* the world and *of* the age, a set of multiple, often superimposed and overwritten, imprints of space and time at their most banal. In turn, sacred writing that arrogates the beginning for itself must sever its connection, indeed its lifeline, to its not-yet-formalized secular counterpart. Styling itself as unworldly, otherworldly, timeless, ageless (*sans* age, universally valid, germane to every age and time), it pushes off from that which will come after it—the secular before the secular. Secularization, then, is not a forward-oriented thrust but one that recaptures the debris that theology and metaphysics, the one as the other, have left behind in the course of their institution.

Whether in its standard variant or in the heterodox installment I have just outlined, the story has a continuation. Just as the postsecular turn has rounded off the tradition of Enlightenment criticism, secular writing has entered a new phase, a new age, another *saeculum* I propose to call *the writing dump*. This phase is complex, if not contradictory:

—on the one hand, we are experiencing an extension of secular writing, hitherto known as literature, on a hyperbolic scale and at an ever-accelerating pace. Everything—every inkling and inclination, reference and reaction—is obsessively committed to writing, often in a highly abbreviated form. No mental artifact has the right to exist without being inscribed, externalized, publicized. The writer, texter, or message sender recognizes herself almost exclusively in the accretion of online profiles, accounts, and digital pages. Our age is absorbed in this feverish virtual accumulation involving the entire world: transmitter and receiver, shared contents and audience. The democratizing mission of literature seems complete.

—on the other hand, we are witnessing a rupture in the history of writing. A quantitative surge triggers a qualitative change. Writing is massified beyond the scope of print and includes instant messaging and online chats, the texts composed and sent ("texted") as instantly deletable on an apparently ephemeral digital substratum, yet archived somewhere on a digital cloud beyond the users' knowledge and reach. Its falling mass is unavailable to careful, fastidious reading practices and, at best, invites browsing. In the writing dump, everyone is a writer, no one a reader: writing exists sui generis, for its own sake, abutting the sacred (as separate) from another direction. Genres are, by and large, irrelevant. *This* world and *this* time of a mutant secularism are the digital world and time of writing itself.

The world and the time we designate by default as ours are not *this* world-time; they are the appendages of an abstract code, the inscription as such. In the *saeculum* of information that is "without age," we have fallen into the writing dump and keep falling, piling up (not least paperless materials), being piled upon, dumped upon. It is, consequently, necessary to take stock of this situation from within, *in medias residuum*.

I would like to bring into relief two veins running through the monolith of the writing dump. The first is the benumbing modeling of existence on an abstraction, with actuality formatted for virtual viewing. (Format is the semblance of form in the era of technicity. *Estamira* sends us a stark reminder of this verisimilitude.) Aided by a welter of abbreviations and by a general unlearning of spelling, the writing dump begins to resemble a pile of symbols, the code at its purest peering through the coded text.[1] A condition of possibility for meaning-making, the code as such is meaningless. Its coalescence with the written stuff fails to reach transparency; quite on the contrary, a pure code drives writing to absurdity, to nonsense that does not breathe with the promise of sense. Everything haphazardly falls into the writing dump and nothing comes out of it, the "feed" indigestible.

The second vein I wish to pursue is made of the shreds of past ideas. For accuracy's sake, let us say that the writing dump does not recycle bodies of thought, incapable of reprocessing that which it receives. (*Word processing* is a misleading idiom that obfuscates word packaging and word stacking.) Bits and pieces of old arguments and concepts are put to work again, with not a smidgen of awareness of their historical context, in another configuration that gives them the air of novelty. In the endless closure of metaphysics, there is an accumulation of disjointed ideational scarps: object-orientation decoupled from the subject; *America First!* insensitive to the deadly precedents of rabid nationalism and to the regional or global interdependence of political decisions on climate, migration, or the economy; "new" materialism and realism fighting little more than a strawman of idealism; appeals to the pioneering spirit of humanity on a space-colonizing mission, clueless about the social and environmental harms of earthly colonialism . . .

A writing machine, the figure of high modernism, no longer fabricates these and many other flotsam ideas. Depersonalized and dehumanized, mechanical intentionality still needed to process raw materials so as to craft end products out of them. The writing dump has none of these qualities, and, with the active receptivity of reading out of the picture, it passively takes over the receptive function, heaping inscriptions up. If there is anything mechanical about writing today, it is just how it drops on us and drops us en masse.

The rapidly falling written mass is an information dump. Deconstructive hesitancy between the unconditionally hospitable institution of literature and the anonymous reception of whatever comes its way careens heavily toward indifference and undifferentiation in the age of the writing dump. This age, this *saeculum*, borders on agelessness: as Shakespeare has taught us, there are neither sequences nor stages in it—nothing but the inertia of mass accruing in the atmosphere of feigned freedom ("anything goes," nothing comes out). Distinctions between *before* and *after*, other *saecula* and the avowedly nonsecular epochs, are inapposite there where epochs and time itself are pulled into our ageless age's homogenizing vortex. A little counterintuitively perhaps, the unsurpassable indeterminacy of the writing dump proves to be the twin of sacred inscription with its delusion of grandeur, eternal applicability and transcendent determinacy.

And it flirts with worldlessness, too, this world—this *saeculum*—unabashedly exaggerating the tendencies of secular writing. What sort of a world is it that winds up in the writing dump, where every mark is as significant as the next, and so utterly insignificant? Can a world survive after its meaning grid has ceased to send flashes, dimming in parts and brightening in parts? What does the trawling net of the writing dump do to the ocean floor of meaning it is tugged along? What does it inflict on the seabed of expression, of language and sense?

That is not to say that *the world* (and, accordingly, worldly-timely—secular—writing) is beyond reproach. As we have seen, its meaning is deeply imbricated with the theological or metaphysical, otherworldly, true reality, which is its starting point and remainder, its impetus and reflex motion. Striking at the unworldly origin of the world, contemporary philosophy splinters it into worlds, whether along the lines of modal logic, which stresses the plurality of "possible worlds," or of environmental phenomenology, which attends to the unique constitution of experience by human and nonhuman beings.[2] Every piece of writing is a convolution of scripts upon scripts, many of them not written by humans. Every *saeculum* is *saecula*, takes *saecula* to unfold, to exhibit the multiple, partially intersecting worlds and times comprising it. Does the wild proliferation of worlds absolve the world?

Its new-fangled spatiotemporal complexity notwithstanding, the world has been a sacrosanct concept, with few notable exceptions.[3] My not-so-modest proposal would be to cease holding onto this chimera, to let it drop together with the other fictions of metaphysics, to allow it to end with the view to preserving the remains of the earth and the other elements.[4] Correlatively, we could imagine the end of sacred *and* secular writing, as we rid ourselves of the *saeculum*, which is but a shadow of the theological and metaphysical constructs of time and space. Perhaps, writing only begins in earnest after the end of secular writing.

The dump unworlds the world; the writing dump makes manifest the movement and state of unworlding. Its truth is in illuminating the current

Figure 17 Drowning in, Anaïs Tondeur, 2018–20, Pigment print on Murakumo paper, 42 × 63 cm

world catastrophe, the catastrophe that *is* the world. The contours of the catastrophe are not entirely negative, though: there might be a still indefinite something that could flourish in the place of the world. In a darkly optimistic twist, the dump as the crucible of our age may turn out to be an unwieldy contrivance for hastening not the advent of an otherworldly true reality but of the other-than-world. Neither sacred nor secular, a new style of writing would be requisite for dwelling there.

parts of the void

The sky and the well. The abysses of the waters above and the waters below have been, since Greek and Hebraic antiquities, the indicators of infinite depth, between which the earthly fold takes its precarious place, between which the place takes place. Existence is an interruption punctuating the void, suddenly cleaving it into parts.

You may think that you are free to choose where to gaze: up or down, at the sky or into the well, at the abyss above or below, to surrender to the lure of transcendence or the enticement of immanence. But that is not quite so. Before you stare into the void, the void reflects itself across a thin partition, the shaky line of existence punctuating it. The void stares at itself (and at you): the water below mirrors the water above that returns the anonymous look. The gaze without anyone gazing travels back and forth, inside and outside us. "We are two chasms—a well staring at the sky [*Somos dois abismos—um poço fitando o céu*]."[1] The arithmetic of the abyss starts at 2.

Could intimacy with the abyss surreptitiously account for the laughter of the Thracian maid at the sight of the philosopher Thales tumbling down into a well? As Plato narrates the story of the pre-Socratic in *Theaetetus*, "While he was studying the stars and looking upwards, he fell into a well [*pesonta eis phrear*], and a witty, graceful Thracian maid jeered at him . . ." (174a). Her *knowing* laughter intimates that she is amused not with the impracticality of the thinker's gaze scanning the skies but with the confirmation of what she has gleaned for herself from the well's unblinking stare—the stare so familiar to her who has come there daily to draw water—namely, that downcast eyes toss you up and looking up plunges you down, into the whirlpool of the void's self-reflection. She is versed in the arithmetic of the abyss, unlike Thales who separates mundane practicalities (say, of walking or business pursuits) from theoretical contemplation, breaking the mirror of the void.

The esoteric message of the story is that the maid is more of a philosopher than the philosopher. He learns the movement of ascending but crushes upon descent; she is aware, as Heraclitus also is, that the way up *is* the path down, even as one climbs it (*hodos anō katō*). Where he sees a straight line leading to heavens, she espies a tangent of a circle that, rotating, recycles desires, aspirations, and the soul's itineraries leading toward their satisfaction. He is dropped by his doctrine and the unfulfillable wish to stay up in the celestial sphere; she is uplifted by her nondoctrine—by her wittiness, according to which the stronger the delusion, the harsher the fall—and glides down in a smooth, controlled descent completing the lower, geotropic part of the circular path. (Being dumped is a traumatic payback for having disavowed not only geotropism but also, in a more encompassing manner, the circular motion, of which geotropism is the descending part.) He is wise—he "displays his wisdom [*epideixin . . . tēs sophias*]," Aristotle says in *Politics* (1259a, 19), yet how wise is it to put one's wisdom on display?—but not discerning (*phronimos*); she is discerning and perhaps wise, too. He studies being; she is initiated into the mysteries of the void, wordlessly, with laughter alone for a sign, or at least for the only sign she gives that makes it into Plato's report of the anecdote.

Scrap "though philosophy begins in wonder, it may end in dread"! Philosophy begins not in wonder but in being dumped.

When the debris storm of world history dumps the sky onto the earth and flings the earth into the sky, confusing the upper and the lower regions, it is crudely parodying the confusion that reigns outside the limited sphere of being. I am characterizing this as a crude parody because it muddies the infinite reflection of water in water. Indeed, in the dump and in sky-well communication, mayhem, disorientation, and bewilderment prevail, albeit for different reasons and with diverging consequences. *Either* we are thrown into a maelstrom that eliminates the distance necessary for minimal discernment, *or* we are caught between two elemental mirrors, each of them sending us to the other and provoking dizziness, as do certain questions lacking an answer. Psychic life thrives on the confusion of separate chasms (Pessoa), not on the unreflective folding of everything into nothing. We feel and think between the well and the sky, which is the conventional figure of the inbetween. And we become insensitive, thoughtless, and unthinking when the above and the below are dumped into one another, the inbetween obliterated.

Slicing the void into parts is dividing the indivisible. (Chopping a living body into pieces is dividing the divisible and the indivisible, the body and its life, which structurally corresponds to the void.) The outcome of the division befits the procedure: the difference between parts of the void is an indifferent one. A difference voided, it is the difference that makes no difference and does not differentiate between one part and the other. In the global dump, parts of being appear and are treated precisely as parts of the void, their differences immaterial. But what parts the void? With what or by what is it subdivided? Does its segmentation delimit it? If yes, is this delimitation a prelude to containment (inhibition, the restraining of negativity) or to the release of the limit from the spell of its limiting effects (negativity running amok, unchecked)? In other words, does the void's delimitation void the limit?

Parts of the void emerge out of the void's parting against itself. Only in sheer nothing is the void one and the same, equal across its imaginary sections, perfectly continuous and contiguous beyond any measures that could ascertain the relations of equality, continuity, and contiguity. Were it subdivided in a daring thought experiment, each of its parts would have accommodated the whole, neither more nor less of a nothing than the next. For, *ceteris paribus*, assuming that all is nothing, a part of all (i.e., not-all) is nothing as well and is, thus, equivalent to all. Our actuality's love affair with the indifference and undifferentiation of the dump strives toward pure nothing also in the way it partitions the void.

Biblical cosmogony presents another alternative. There, the act of creation delimits the void: "And God made the expanse [*raqia*], and divided [*va'yavdel*] the waters which were under the expanse from the waters which were above the expanse" (Gen. 1:7). Logically, the flood is a harbinger of decreation: it threatens to restore the original confusion of the waters above and below, to pour the sky out into the well. The expanse, that which spreads out, *raqia* in the sense of *the extended*, is the partition that contains parts of the void, assigns to them their proper places and guards them there. It does not cut its ties to the watery abyss, but merely arranges aquatic delimitations along a vertical axis, keeping open the option of undoing the limit in the future.

The same goes for the chaos of the earth that "in the beginning" was *tohu va'bohu* (Gen. 1:2), "formless and void," dreary and barren, a wasteland of desolation, a desert-dump. (Luther rendered this expression in German as *wüst und leer*, desolate and vacant; Franz Rosenzweig favors *Wirrnis und Wüste*, confusion and a desert.[2]) The elements teeter on the verge of nonbeing when they are not yet, or already not, articulated among themselves. Water and the earth are the noncommunicating parts of the void, so long as they persist apart from each other. Darkness (*khoshekh*) is a sign that the desert-dump is isolated from fire and its luminous power, which will come into being in the biblical verse that follows. This isolation allows the desert-dump to skirt exposure, to maintain itself in reserve, concealed.

Like water that may revert to the elemental void, which preceded the separation of creation into the upper and the lower regions, the earth, too, may backslide into the chaos of the beginning. Jeremiah prophetically depicts these eventualities, painting the landscape of exile with the "environmental" palette: "And I looked at the earth, and here it was formless and void [*tohu va'bohu*], and at the sky, and it had no light. I looked at the mountains, and here they trembled, and all the hills swayed. I looked, and here there was no man, and all the birds of the sky fled" (Jer. 4:23–5).[3] Undoing the work of creation is either fusing all the elements or segregating them (the earth from the sky; the earth and the sky from light), circumscribing them to their idiosyncratic, idiotic domains, and plunging them back into the void. The void is the aftermath of parts falling apart, unmooring from one another, expunging all traces of their articulations in a world, neglecting their partiality. Rather than parts of the void, the discombobulated elements come down to the void of parts, the void reawakening in those parts of being that, having been compartmentalized, lose contact with their counterparts.

Tohu va'bohu captures the sense of disarray as it awaits the pulling-together of the world, the becoming-world of the world in creation. Between "dreariness" and "barrenness," the Hebrew word or the letter *and* (*va*, *vav*) pre-articulates the two names for elemental disarticulation. In Midrashic commentary, notably in R. Abraham bar Hiyya's *Hegyon ha-Nefesh ha-'Atzuva* (*Meditations of a Sad Soul*), *tohu* is understood as matter and *bohu* as form. Relying heavily on Aristotelian hylomorphism, bar Hiyya writes: "If you compare the explanation of *hulē*, of which it has been said that it has no image and no form and which cannot subsist by itself, to *tohu*, you will find that they are the same thing . . . *Bohu* is the form that covers *tohu* and sustains it" (2b–3a).[4]

The formal and material aspects of the void are its *ur*-parts, presupposed by all future divisions and partitions. Their linkage through the word *and* tempers the void's desolate confusion. When parts of the void open unto other parts, they void the void, internally transmuting it into something else. Their pre-articulation is the very thing that keeps alive the hope for overcoming the unworlding disaster. If I hyphenate the desert and the dump, omitting the connective *and*, I do so to emphasize the dearth of actual and potential articulations in the global desert-dump, let alone the articulations of potentiality and actuality themselves.

Our void is void of meaning and of the future. It is a dump for matter and form, potentiality and actuality, the parts rattling in it side-by-side, neither combined with nor quarantined from one another. On the one front of ontological warfare, we are besieged by empty virtual irrealities, possibilities allergic to actualization, and, on the other, by the lingering leftovers of realized projects, actualities that would not ebb away. In the void, the possible is inexhaustible, despite the sense that real possibilities, those that can be carried through to fruition, have been depleted. We are stuck at square one, back at the confused beginning that tolerates everything, save for its own elapsing. Parts of the void are, essentially, possibilities barring actualization and indistinguishable from the impossible. They are the differences that make no difference and that are indifferently hoarded in a deserted archive.

When a constellation of ruthless forces dissimulates itself behind a façade of freedom and pure possibility, human action is deprived of frictions, temporal developments, rites of passage and stages of unfolding that are the outcomes of negotiating its envisioned course with others and with the world. The virtuality of the void foists upon us a beginning that never ends, and, therefore, never really begins. It is the desolation of first givenness that blocks the second. Virtual irreality parodies labor, which, in Arendt's philosophical scheme, is "caught in the cyclical movement of the body's life process, [and] has neither beginning nor end,"[5] as opposed to work marked by "a definite beginning and a definite, predictable end."[6] The dump implements labor without work that, by stymying the capacity to begin, proscribes politics as "the capacity of beginning something anew, that is, of acting"[7] together with others. Ergo, it not only creates a problem, but also invalidates in advance viable solutions to the problem, locking us up in a labyrinth of infinite possibilities and throwing away the key.

The other front, the other side of the ontological siege, under which we live-die, consists of artifacts that have perversely given a body to our metaphysical hankering for eternity: microplastics, depleted uranium, increasing concentrations of CO_2 in the atmosphere, and so forth. If they have been turning up, time and again, on several pages of this book, that is because they have never really disappeared; they are here to stay, dragging the *here* into the void. We can do no more than turn our backs on these by-products and avert our eyes for a while, and even then our peripheral vision will be overcrowded with them. But why should they resurface and come into focus now? How are they relevant to the void?

The void is stitched with the double thread of nonrealizable possibilities and leftovers from realized projects that simply would not go away. These apparently incompatible strands are the afterglow of an event in being that is comparable to nuclear fission: the splitting of potentiality and actuality. Unplugged from the energy supply that is their synergy, parts of the void become the impediments to becoming, potentialities declining the invitation to come into being and actualities reluctant to slip out of it. *No, no*: echoing one another by turns, resonating in a dry well—unless it is the sky that has dried up—the two negations have otherwise long ceased communicating. Their medium and message is a desert-dump, the desert of dry possibilities obstructing every passage to actuality and the dump of petrified actualities that are rapidly colonizing space and time.

The resonance of the two *no*'s in the desert-dump channels a decisive and unmitigated contradiction of the present: the virtualization of currencies, education, and interpersonal relations is dogged by obstinate reliance on high-carbon and nuclear energy. So, the sum total of energy that goes into individual transactions with bitcoin, a leading cryptocurrency, surpasses the annual electricity consumption of Ireland.[8] Compounded by an addiction to fossil fuels, efforts at reviving the nineteenth-century vision of a sovereign and self-contained nation in the United States are, rather than anachronisms, the powers of the dump complementing the virtual desert of the Silicon Valley. Wildly disparate as they seem, these, too, are parts of the void—of the same void—as evinced, precisely, by their disarticulated coincidence.

*

A well is a pit in the earth, dug up until the shovel meets the waters below. It burrows in, courses down through one element so as to occasion other elemental encounters: of waters with waters, of the sky with what lies belowground. To give these encounters a chance, the earth makes room at the price of its own uninterrupted presence. Its partial withdrawal creates determinate, delimited emptiness, rounded off at its edges, a peephole for an infinite speculative reflection of the same at a distance from itself. The well and the sky practice elemental, anonymous attention, a faculty more ontological than psychic, one that Simone Weil associates with vacancy and retreat: "Attention consists of suspending our thought, leaving it detached, empty, and ready to be penetrated by the object."[9] Attention as voiding, within limits.

In the desert-dump, the well runs dry and, abandoned, no longer gathers the elements around itself. But the void that was at its heart stays. Vast in its devastation, it detaches from the circular rim and, in a monstrous hypostasis, proclaims itself absolute. Without the roominess of a delimited opening, parts that were to communicate with one another are crammed into a heap at random. There are more of these parts than ever before and no more parts: they massively multiply and refuse to participate in anything beyond themselves. Should we be surprised? A dumped heap is, after all, the desert void.

in-formation

Information is a promising word. It is a pledge, in fact, to combine matter and form in a single word. On its semantic surface, it says that the stuff of the world reaches us in a particular form and that, once we welcome these givens, these data, with our senses and minds, they have a formative influence on us: we ourselves (our views, opinions, decisions) are informed. This positive dimension of information coexists with another, also integral to the word, namely the negation of formation. Information arrives unshaped, lacking in form, a flood of data that does not move in manageable streams. It buries or inundates its transmission lines and lineages that could act as ersatz traditions. In some sense, it is the dump in a nutshell, with its mass of rapidly dropped data spilling over and washing away the perceptual thresholds, cognitive schemas, attention spans, and other finite capacities of its recipients.

Mutually exclusive senses of information obtrude upon us simultaneously. It gains on us along both dimensions, along the two colliding tectonic plates, at once, informing and deforming, transmitted and dumped. Outside its circuits (and we should not be callow: it is impossible to actually berth at that exteriority) we are uninformed, ignorant even; within (and, as we shall see, the world interior of information[1] suspends many of our received ideas regarding inwardness), we are swamped, knocked off our feet, coming undone by its surfeit. Whether spatially or temporally, it disorients us, and yet no system of orientation could be independent of it.

When the Information Age was still unimaginable, Thomas Aquinas contemplated *informitas materiae*, the "unformedness of matter" at the time of creation, in what is no doubt a reductive take on *tohu va'bohu*. In this prehistory of information, in this obscure portion of its genealogy, Aquinas references "various saints" in writing that "matter was in a certain manner unformed, and afterwards it was formed; also time was in a certain manner unformed, and later formed and distinguished by day and night [*materia erat quodammodo informis, et postea fuit formata; ita tempus quodammodo fuit informe, et postmodum formatum, et distinctum per diem et noctem*]" (*Sum.* Ia, 66, 4). Our sick fixation on the possible, the fixation that detains us indefinitely at the threshold of an unending beginning, is at home in the unshapeliness of matter conveyed as information. This unshapeliness is none other than the nonappearance of information, compared to the data it carries. The source of the "givens" is itself not given. Information amplifies the nonphenomenality of the dump, just as beauty, conversely, intensifies the shimmering of phenomena—the shining forth of *kosmos* in and through their shapeliness or formedness.

Seeing that information is both in a form and unformed, in it, merely formal possibilities immediately coincide with formless matter. This contradictory coincidence is emphatically not dialectical: it does not occasion the mutual alterations and mediations of becoming instigated by matter's self-negations. Matter is bogged down in mass, the massiveness of the unformed and perhaps the unformatted, where the materiality of different bodies is one and the same dump. Aquinas again: "the unformed matter of all the bodies is one [*una sit materia informis omnium corporalium*]" (*Sum.* Ia, 66, 2). What better way to describe the world from the standpoint of information?

Neither diurnal nor nocturnal, the unformedness of matter in Aquinas is unlight, not to be mistaken for darkness, and untime, not to be conflated with eternity. His depiction is better attuned to the dump than Plotinus's "dark matter" and Paul's "enigmatic vision"; it conveys the sense of unbeing we have seen Heidegger detect in Heraclitus on *sarma*. Unformed matter emerges before the distinction between day and night not only because light has not yet been created at that stage in the biblical narrative but also because there is no one to whom this distinction would make sense. For the same reason, Aquinas ties nocturnal-diurnal difference to time. A cadence of cyclical alterations, a moving pattern of celestial events, finds its meager likeness here-below in the paces of finite existence—for instance, the circadian rhythms of various lifeforms. With the intensity of artificial lighting that sets urban skylines ablaze at night, temporal difference dissipates, formed (temporalized) matter drifting back into undifferentiation. Decreation all over again . . . The city that never sleeps is soaked with information, so much so that it fuses with the stuff that fills it, formless and in an inflexible formation. Circadian rhythms in humans, birds, and plants suffer a severe disruption.[2] The information retinal cells and other kinds of photoreceptors register is always that of radiant energy: it is daytime all the time, irrespective of the time of day or night.

Ever an aficionado of double frames and an exponent of the protodialectical coimplication of opposites, Augustine, whom Aquinas cites and with whom he takes issue in his *Summa theologiae*, distributes *materia informis* between the spheres above and below, *superiorem atque inferiorem*. Augustine's provocative thesis is that "each unformed matter is of heaven and of the earth, namely spiritual life as being possible in itself, not turned toward the Creator . . ., and corporeal life, which may be understood through the privation of all corporeal qualities that appear in formed matter [*An utriusque informis materia dicta est coelum et terra: spiritalis videlicet vita, sicuti esse potest in se, non conversa ad Creatorem . . .; corporalis autem si possit intellegi per privationem omnis corporeae qualitatis, quae apparet in materia formata*]" (*Gn. litt.* I.1.2).

Information materializes when the unformed matter of heavens crashes into that of the earth and virtual being possible in itself (*esse potest in se*) meets, head-on, that which is defined by the lack of all corporeal qualities (*per privationem omnis corporeae qualitatis*). Kant would have stressed the categorial deficiencies in such binate provenance: from above, unformed matter is in cahoots with a lopsided modality, where possibility spurns actual existence; from below, it privileges the category of quantity over quality. This is the tenor of information. In its pure, purified, distilled state, information transmits quantitative data and virtual possibility, with quality and actuality shaved off as so many annoying interferences with the formal transmission process. We are supposed to be informed by a categorially impoverished skeleton of the real, the quintessence of *what is*, the DNA of being, its basic code. To extract the valuable nucleus, we are urged to discard that which is viewed as no more than the outer trappings of existence: its phenomenality. Once the exercise in reduction is complete, a colossal abstract form is mounted on feet of clay, shapeless and unreal. Hence, the paradox of information that is in a form and unformed.

To Augustine, the phrase "And darkness was over the face of the deep" in Genesis 1:2 signifies an unformed or deformed life disconnected from its source and incapable of self-actualization: "The translation of the expression the *dark deep* refers to the nature of unformed lives that are in this condition when not turned to the Creator, which is the only way life may attain form and not be the deep [*ut translato verbo tenebrosam abyssum intellegamus naturam vitae informem, nisi convertatur ad Creatorem: quo solo modo formari potest, ut non sit abyssus*]" (*Gn. litt.* I.1.3). While his initial reflection touched on the quality-free character of information, Augustine's reading of the second verse of Genesis helps us grasp its nonactuality.

The dark abyss of information (or unformation) is the out-turn of turning; a chasm opens as a function of the direction one faces or doesn't face. Like "the nature of unformed lives" (*natura vitae informes*) in Augustine, information revels in itself—in its chains, data banks and codes, circuits, lines of transmission and propagation, speeding along a highway to nowhere. Whereas the ever-dwindling territories lying outside its purview are perceived as an abyss of the unknown, the abyss is the empire of information (or unformation) itself. What applications it may and does have, it is ultimately a possibility in itself, virtual, dormant in its autistic enclosure, turned to itself alone. It stands for the impossibility of receiving an actual form by negating itself (in other words, by negating its *in itself*) and for the impossibility, also, of navigating and negotiating between itself and the other, something that, in the first instance, would require realizing the possibility and the necessity of turning toward the other by means other than translating alterity into data. In the deluge of information, in this frozen flood, nothing stirs, nothing turns (and time least of all), nothing comes to be.

As it forms (or unforms) the world in its image, information does not hand itself over to modifications by that world and, as a result, abides in its formlessness. Herein is its power, potency, potentiality, or possibility (*potentia*, *potestas*) hardened against actualization. Information is extracted from the core of the entities presumed to be its vessels. But its own *in itself* is dubious: it has no interiority. Essentially, information is beholden to the outside insofar as it enucleates the meaning of beings it reduces to a code and insofar as it spills over every receptacle, into which it is poured. This feature complicates to the *n*th degree our relation to the digital cage that is the information grid. If a passage from captivity to freedom typically implies breaking out of a confining stricture, our detention in the *in itself* of information is an arrest in that which lacks inwardness. When one is beset by a flurry of pure possibilities, totally on the outside "within," where does one look for an emergency exit? Good fiction writers are abreast of the fact that the information dump weighs heavily on plot development and the building of narratives. A massive, growing pile of data asphyxiates discourse, articulation, and even words, among other interpretations of *logos*. Now, the foremost obstacle to plot development emanates not, as one may expect, from actuality, which would clip the wings of imagination, but from possibility severed from other modal categories. Where form is only formal, as in the shapeless pile of information, anything is possible, which is to say that nothing is *actually* possible. The dump unhinges the double frame of actuality and possibility, such that the latter dodges concretization and the former is excised from the flows of time. It explodes the last footbridge connecting movement and rest.

It is a fallacy, and one propitious to the dump at that, to regard development as a dangerously teleological anachronism. Time trajectories are winding and convoluted, detours upon detours, harking back and leaping forward, jumping ahead *by* going behind, regressing *by* rushing forth. Along their zigzagging, looping, uncertain tracks something unfolds, emerges, forms, deforms, and reforms, something comes forth into a fragile and fleeting presence. Information, however, is the realm of untime, which goes hand-in-hand with the dump's unbeing. On its own terms, it has no history, no temporal ordering that would be meaningful within its immense outpouring. Dropping on us all at once, it muffles the rhythmicality of time in the white noise of space, causing us to drift back to the condition of James's infants assailed by all the senses and stimuli simultaneously. So much so that, rather the information technology itself, the lives and preferences of its human users deducible from their online activities are (in digitalspeak) "data exhaust," valuable detritus for megacompanies like Google to exploit and profit from.[3]

To managerial logistics, the untime of information is alluring as a means of gaining control over time. The control methods looming over us today do not assume the shape of a tremendous, fine-tuned, centrally run machine. They consist in part of a nanopolitical,[4] personalized, wired, information-based set of psychodigital manipulation techniques that fit each subject as a glove, if not as the skin of the hand sheathed in the glove. Their other part is impersonal surveillance by Big Data, which is to say without anyone in particular doing the work of digital spying. The ones under surveillance—all of us—are not individuated by the relentless techno-gaze that is as indifferent and undifferentiated as the effects of toxicity. Such diffuse surveillance corresponds to a feeling of never being alone *and* being more alone than ever. Control unyoked from a determinate relation to the controlled becomes uncontrollable, again in parallel to the unleashing of a toxic flood onto the world. Plus, as Donald Trump has intuited, the cutting-edge tool enabling one to achieve control is chaos, or a dump of causes and effects stacked with no rhyme or reason, let alone distinctions between micro- and macro-levels of power. Statistical facts are no less culpable than their "alternative" cousins: impersonal technocracy and a capricious political leader play on the same team. Swearing by the information dump is only nominally opposed to thumbing one's nose at hard data.

There is a price to pay for the political uses of chaos, to be sure. As I have just mentioned, the latest control apparatus is itself uncontrollable, if dazzlingly effective. To enervate, often by means of dispiriting references to a mindboggling complexity of the world, the articulations of what is afoot, is to deny imagination the luxury of conjuring another path or mode of unfolding. To withdraw the ground of saying is to invalidate, in advance, all countersayings and stated contradictions. Under the neutral camouflage of the information dump where anything goes, nothing is actually going on. Things stagnate beyond the stand-off of time and eternity. Untime preserves the otherwise unsustainable status quo.

An associated technique of time control is discernible in the computational field under the revealing name of *memory dump*. This is the procedure of extracting the totality of information from a computer's RAM (Random Access Memory) and saving it on an external storage drive. Memory dump functions as a diagnostic measure in the situations of application or system failure. The procedure provides a panoramic overview of all programs and applications just before their termination. In a memory dump, time stops retroactively: the constellations of information (memory locations, program counters, etc.) appear as they have been on the brink of a disaster. The technological hope is that, come what may, information would be recoverable in toto, preserved on another disk or device. It confirms that matter and time are inessential to information, that the interplays, errancy, eventfully transitory and surprisingly leaping or looping movements of moments are dispensable as far as the computing process, which dumps the contents of every single instant into digital space, is concerned (or unconcerned).

If the structure of our consciousness is cybernetic, if it lends itself to analysis into bits of information, then a computing memory dump will furnish the model for a mind dump, an overall digital schema of memory and cognition prepared for posthumous preservation. The advocates of this method mistake information's untime for a gateway to eternity. They reckon it enough to save (note the salvific overtones of computing in this locution) the whole by copying data indiscriminately, indifferently, by means of dumping, with the view to resynthesizing that which has been saved at another time and place, at *any* time and place, virtual or "real," exactly as it had been.

Lost in digital translation are the singular interactions of the mind with its environment that transpire on the margins between the two, where consciousness abides outside the screenshots of a memory dump. Nor does embodiment lend itself to operationalizing on the foundations of informatics: a living body is not a piece of hardware to cast off and replace with another upon a safe replication of the software. So-called Whole Brain Emulation and mind uploading will have retrieved nothing more and nothing less than what the computational paradigm has deposited there, far removed from experience, from a (by turns) active and passive undergoing with the other at one's limits. They will have preserved formless information stuffed into the empty drawer of consciousness.

rameau's nephew for the twenty-first century

We've taken note of how in the ancient worldview tiny parts of vast cosmic elements are circumscribed and temporarily detained in varying proportions in bodies, animate and inanimate. This is not the sole mode of interrelating the micro- and macro-levels of existence in Greek thought. In Plato, certain psychic structures and faculties are strictly analogous to those of political life, so that the "letters of justice" inscribed in the soul are written in all caps or in a different font size in the city. "Since we are not clever," Socrates says in *The Republic*, since we do not have enough discernment "to read small letters from afar," we should examine "the same letters . . . but larger and in a larger place," the place of the *polis* (368c–d).

For our purposes, Socratic words mean that the mind dump reverberates on a grand scale in politics. A haphazard and inconsistent heap of character traits—those of the new leaders and of their followers—finds a perfect match in populist movements, thick with self-contradictory promises. Attuned to the vacillating moods of the masses, populism caters to indignation with the status quo and anger toward immigrants, refugees, and other foreigners. Responding to a thirst for decisive action, it leads to entrapment at the hands of the worst demagogues (the politics of antipolitics, so to say). Nostalgia for bygone national or imperial glory and demands to modernize are mixed in it pell-mell. The scrambled letters of injustice are indeed the same (save for their respective font sizes) in our polities and the "personalities" of their leaders.

The dump we are mired in is not only corporeal-environmental but also psychopolitical. Its logistics expose the fault lines separating twentieth-century fascism from the more recent iterations: while classical fascism was a politics of the masses molded and elevated to the totality of state form, its parodic version today is the politics of the masses placated in the disarray of their conflictual aspirations and manipulated in their formlessness. The masses and their populist leaders fall together, dropping both political discourse and the tactics to garner votes below the threshold of propaganda. The ensuing pile of proposals and assertions is no more than a heap of soundbites, tweets, and memes. Matter drained of its inherent forms is, after all, tethered to pure potentiality, ever ready to ripple whichever way.

Denis Diderot's 1774 dialogue *Rameau's Nephew* is remarkably apropos to today's political situation. The work's eponymous character is by no means a politician, but his psychological profile is a dump of qualities uncannily reminiscent of our populist leaders, chief among them Donald Trump. "The concepts of honour and dishonour must surely be strangely jumbled in his head," Diderot writes about him, "for he makes no parade of the good qualities which nature has given him, and, for the bad, evinces no shame." "Nothing could be more unlike him than he himself is [*Rien ne dissemble plus de lui que lui-même*]."[1] Often, he produces but "a tangle of words," *entortillage*.[2] In an exaggeratedly Machiavellian vein, he believes that "nothing is more useful to the common people than lies; nothing more harmful than the truth."[3] "I am a man of no consequence [*Moi, je suis sans conséquence*]," he says about himself.[4] Wealth further sustains this position: "If you are rich, no matter what you do, you can't be disgraced."[5] Freedom of speech he interprets as speech free of thinking: "I make good use of my freedom of speech. Never once in my life have I thought before speaking, while speaking, or after speaking."[6]

There are many more similarities between Rameau's nephew and Trump yet to be brought up than those I have quoted. A close reading of Diderot's masterpiece will reward whomever undertakes it with a nuanced view of the psychopolitical dump. We will shortly get back to the details of this illuminating text, but not before commenting on the role of potentiality in populism.

Formlessness and indeterminacy, the one boosting the other, contribute a lion's share to the mishmash of personal and political qualities. Enhanced by the flood of online information, these are the features of pure possibility, its development toward actual and palpable outcomes arrested. For Arendt, constant change is a distinguishing trait of totalitarian power, which, if I might add, approximates power in its conceptual ideality, the potency of *potestas*, a sheer potentiality oblivious to that for the sake of which it may be wielded.

In totalitarian administrations, Arendt perspicaciously observes, "constant removal, demotion, and promotion make reliable teamwork impossible and prevent the development of experience."[7] Here, too, we cannot help but think of the perpetual rearrangement of Trump's cabinet since the start of his term in office in January 2017. Paralysis in the first beginning that distinguishes dump politics is dissimulated in unending movement. Arendt, for her part, attends to the phenomenological, psychological, and ontological ramifications of the totalitarian flux in "the development of experience," that is, on the psychic side of the psychopolitical complex. She hints that totalitarianism, much like the dump in our account, interferes with experience itself, and, most crucially, with the experience *of* itself, with the possibility of registering a rigorously totalitarian experience.

In contrast to ancient psychology, which had always bordered on biology and zoology in the ample sense of a discourse of and about life, modern psychology has been immersed in the study of potential dispositions, penchants, and personality traits. Already Hegel has acknowledged that modern psychology's object of knowledge is a dump: "Psychology must at least wonder that in spirit so many sorts of contingent things or so many heterogeneous kinds [*so vielerlei und solche heterogene einander zufällige Dinge*] can exist alongside one another as in a sack [*wie in einem Sacke*], especially since they show themselves not as dead things at rest but instead as restless movements [*als unruhige Bewegungen*]" (*PhG* §303). Rameau's nephew is the psychological archetype of modernity: the "restless movements" of his inconsistent personality are contingently thrown together as in a sack. More precisely, he is the psychological *type* of dump mentality, with Trump for its psychopolitical twin.

The type (*espèce*) Rameau's nephew attributes to himself represents averageness. It is not a specific kind, but type in general, the vacuous type of type. People who are "equally pathetic in good deeds and evil" are "what we call 'types',," he declares, which "of all epithets is the most to be feared, because it indicates mediocrity, and the ultimate in contempt."[8] Incidentally, this is one of the sentences from Diderot's work Hegel quotes in *Phenomenology*, so as to distinguish the mediocre type from a kind (*Art*) where the real seems to epitomize the ideal and where "what is *meant* counts for what is [*das* Gemeinte *gilt darum für das, was es ist*]" (*PhG* §488).

In French, however, *espèce* also has pejorative connotations, to which Diderot evidently alludes. Addressing a person one despises, one can cry out: *Espèce de brute!* ("You brute!") or *Espèce d'idiot!* ("You idiot!"). It is the lowest type (a kind of lowlife; hence, the idiom's biological slant—*espèce* is literally "species"—excluding the addressed subject from the cultural domain) *and* the most average, mediocre kind. The same element that warrants a typological system of classifications, levels down the categories and gradations within that system, bringing it to naught, devolving it to a dump. If Rameau's nephew and Trump are synonymous with *type*, if they typify it, this type, that is because the only thing that stands out about their ambivalent, volatile, appetite-driven personalities is how well they represent general mediocrity.

Exceptional, exemplary averageness shorn of any adornments and subterfuges provides the psychopolitical dump with a solid foundation. In his rumbling, free-associating speeches, Trump voices some of the darkest desires of the unconscious; Rameau's nephew shamelessly confesses to manipulating and exploiting others. "There was, in what he was saying," comments Diderot, "much that we all think, and by which we guide our behaviour, but do not actually say. In truth, this was the most striking difference between my man and the majority of other people."[9] Both characters give free rein to the destructive, uncathected forces of the id, no longer held in check by the superego. The masses identify with such personalities at the unconscious level, their most typical repressed affects, impulses, and inclinations discernible in the "dumpiness" of the populist leader.

"He"—Rameau's nephew, referred to as *LUI*, or *HIM*, in the dialogue; Trump; a populist leader: the gendered personal pronoun is purposefully deployed here—is not an individual. "He" is, in himself, in his exceptional averageness, the massively piled "they," a mode of being-in-the-world Heidegger terms *das Man*. The mechanisms of identification work to bridge the initial gap between the identifying and the identified-with. Nevertheless, as with biomass and the rest of dump dynamics, the distance between a populist leader and the *populus* is an optical illusion to begin with. His mind dump, a bunch of "ideas all scrambled together [*pêle-mêle*],"[10] is already theirs; their erratic conduct is already his ("ME: What have you been up to? HIM: What you, I, and everyone else is up to: good and bad, and nothing at all."[11]) That is why whatever the despicable acts Trump has committed or may yet commit, they will not in the least hurt his populist appeal. In the dump's disorienting milieu, falling is rising, including in approval ratings. "HIM: When I hear some discreditable detail about their [an exceptional person's] private life, I'm delighted. It brings us closer. I find my mediocrity easier to bear."[12]

The exemplary typicality of a populist replaces the charisma of a fascist leader. Instead of haughty rhetoric, intended to uplift the people, today's massively falling political discourse is peppered with crass observations, incendiary provocations, professions of ignorance and claims to being a genius. "I don't know history," quips Rameau's nephew, "because I don't know anything. Devil take me if I've ever learnt a thing—and if, because I've never learnt a thing, I am any worse off."[13] He suggests that geniuses be "flung into the river with the rubbish,"[14] despite wishing to be a genius himself ("I remember a time," his interlocutor says, "when you despaired at being only an ordinary man"[15]). Irreconcilable possibilities are equally tantalizing due to a pure potentiality endemic to the psychopolitical dump. As Hegel has it, personality features are left indeterminate, latent, malleable into their negations, unless they attain actuality through habitual actions. The condition of the voting masses is likewise vague and uncertain when it is the unconscious, sparked by populist threats and promises, that speaks at the polling station.

Unalloyed possibility, kept alive in the flurry of demotions and promotions that Arendt critically targets in her discussion of totalitarianism, admits clashing courses of action undertaken as though nothing has really ensued. Therefore, the reproach that leans on the inconsistency of the psychopolitical dump is misplaced. The interlocutor of Rameau's nephew tells him that "you have not developed this prized unity of character. At times, you seem to vacillate in your principles,"[16] a message one could reasonably convey to the forty-fifth American president. It escapes the author of the reprimand that there is no such thing as a unity of character outside the unity of action actualizing it, putting its chaotic bundle of contradictory possibilities to work and practically weeding some of them out. "Principles" are but a loose agglomeration of words cut off from deeds and decking out the dump.

Behind a barrage of speeches and tweets, "things get done," often to the detriment of the environment, natural or social, and to the detriment of the populist support base, too. But "getting things done" is not arriving at an outcome that actualizes a given potentiality. Least of all is it the coming to fruition of political action. It would be cogent to argue that the dump is a simulacrum of ideology, interring under its massive pile the true motivations for the things that do get "done." At the same time, the need—previously satisfied by ideology—for a coherent official narrative has become obsolete, which is why public discourse in the United States and elsewhere is sinking to the depths of inarticulateness, its vulgar protagonists stooping lower and lower beneath the standards for decency and taking leave of public mores and good manners.

Psychopolitical disorientation affects spatial consciousness, too, warping the sense of what is above and below. His averageness and mediocrity notwithstanding, Rameau's nephew announces that he dwells in the proto-Nietzschean solitude of high mountain peaks: "For my part, I'm unable to see anything from that tremendous height where everything looks the same—the man with shears pruning a tree, and the caterpillar gnawing one of its leaves, so that all you see are two different insects, each doing its job."[17] The spectator, who, among other things, impersonates a detached scientific attitude, is indifferent to actual differences (say, between a human and a caterpillar) presenting themselves to a gaze lost among pure possibilities. He grounds his position of "tremendous height" on this abstraction that elevates the psychopolitical dump even in the midst of the fall. The greatest obfuscation is a product of translucency, of the transparency of the possible unsoiled by the real.

The vacuity of abstractions, the sloppy composite of "personality" (private, public, TV . . .), and a no less slipshod notion of what people ("they," the they, *das Man* . . .) want, make everything look the same, as though seen from above, following the lead of Diderot's grotesque character. Rameau's nephew formulates equalization achieved through dumping by adverting to the obscene, excremental meaning of the word: "it's all the same. The important thing is to open your bowels easily, freely, enjoyably, copiously, every evening; *o stercus pretiosum!* That's the sum total of life in every condition [*Voilà le grand résultat de la vie dans tous les états*]."[18] Every type of life, every kind of existence, disbands in abundant shit that covers over their qualitative dissimilarities. One more time, the high and the low fuse, including at the level of language: the bombastic Latin exclamation *o stercus pretiosum!* means *O, precious dung!* The dump is not a marginal by-product of life activity; it is a great result of life, *le grand résultat de la vie*, foiling, at the extreme, all other outcomes.

In a further twist, the French text of the dialogue carries a subtle political message. According to Rameau's nephew, copious excrement is the result of life in all the estates, *dans tous les états*, in all socioeconomic classes and strata. Unlike the token Institution of the Estates General (*États-Généraux*), it really mucks up the stratification of society and is conducive to political leveling. It is here that the populist solution to the problem of inequality diverges from the communist proposal. Populism diffuses substantial unevenness in loads of shit; communism nurtures the differences in ability, in what each can contribute to the common good, in the fruit each life may bear before it is digested and expelled by the body politic.

dump philosophy, or the task of thinking in the age of dumping

Dump philosophy? Is this an injunction, to be reinforced with an exclamation mark: *Dump Philosophy!*? Or a description of what happens to philosophy in a dump? Or an assurance: Holding you by the hand, I will walk you through the philosophy of a dump? Or, perhaps, a more sinister commitment to a philosophy *for* the dump?

The last pair of options are of special significance. A cardinal decision is whether, in view of the psychological identifications we have tracked earlier, a given system of thought elaborates a philosophy *of* or *for* the dump, whether it tries to sort out or contributes to muddling the contemporary meaning of being. Standing *for* it, you may be voluntarily or involuntarily complicit: voluntarily, if you justify the growth of the indiscriminately piled up; involuntarily, if you obdurately cling to the idea of purity—be it conceptual or ecological—and refuse to face up to irreversible mutations in the elemental, corporeal, psychic, political, and other scrambled strata of existence. The only feasible resistance to the dump arises from within and traverses it in a countermovement defying its anti-energy.

A thinking countermovement that immanently undermines Western metaphysics has been the prerogative of deconstruction. The issue has never been dumping philosophy as a whole, nor even in its limited capacity as the metaphysics of presence. Derrida knew full well that, at the (endless) end of the day, the battle cry *Dump metaphysics!* and the metaphysical dump it overtly rebuffed amounted to one and the same thing. For one, an outright rejection of metaphysics is the highest form of metaphysics, a thesis empirically verified by the proliferation of "neutral" technocracy, by recalcitrant claims to scientific objectivity, new realisms, materialisms, and other -isms that claim to be non- or postmetaphysical. For another, to throw metaphysics out, to make it fall abruptly and haphazardly, to dump it wholesale, casually and nonselectively, is to engage in its signature activity. To do so is merely to offload everything metaphysics has wrought as collateral damage in its war on finite existence: all the nonbiodegradable waste and flaming biomass that we generate and that we are.

The indifference of metaphysical history, in the course of which spirit dumps matter, persists in the dumping of metaphysics that shatters previous hierarchies at the end of this history. We nonetheless get the impression of a radical break when we must make an *either-or* choice between hierarchical organization and the unimpeded profusion of difference, or, *mutatis mutandis*, between idealism and materialism, the dumping and the dumped. Facilitating this binary decision is the withdrawal of the third alternative: a discovery of meaningfulness in difference, a dynamic *order* based on differences without the violent imposition of hierarchy and its apparently more benign modern versions (rankings, systematizations, classifications). The polarities of order or chaos, a fixed sense or absurdity, the iron fist of unquestionable authority or anarchy are themselves posited on the caved-in grounds of the dump.

Despite its usual association with a rigid structuration of the world, metaphysical ontology comes into its own, its identifying features thrown into sharp relief, in the global dump. It finally succeeds in trashing existence in actuality and without exception. The razing of axiological ladders, the assertion that everything is of equal value in several strains of contemporary thought, implies that everything is equally valueless, that everything (and this *everything* includes *everyone*) shares the status of garbage. It proves, if anything, that not all forms of equality are equally desirable. The permutation of equality that neglects, does not care for, or declines to engage with differences derives from the dump. Spirit is no longer above dumped matter situated below; nothing rises in the equality of the dumped, save for the pile itself. It is this indiscriminate pile, the dump, that towers over the dumped, their formal equality constrained by the not-quite-whole infinitely surpassing the shredded parts. The moment thinking gives its seal of approval to and, at times, glamorizes this state of affairs, it abdicates its ethical, epistemic, and ontological responsibilities.

The metaphysics that is being dumped at present extends beyond modes of cogitation, frameworks of religious or ideological belief, individual and collective psychologies. Its massiveness, the enormity of its throw, does not spare Hegelian objective spirit—political institutions and economic channels, juridical apparatuses and the tissue of civil society. Accordingly, the gesture limited to brushing off the subjective expressions of metaphysical thought in favor of more inclusive, less anthropocentric or hierarchical ways of thinking, is insufficient and redolent of idealism. A social, juridical, economic, and political overhaul must accompany this rejection, and even then our resounding *no* to metaphysics will be largely inaudible in the heap of spent nuclear rods, plastics, atmospheric CO_2 emissions, among the many instances of unwasting waste spawned by "objective spirit" and its cumulative long-term effects.

In the age of global dumping, the Romantics' wistful wandering among the ruins is not a viable itinerary to follow. Rather than ruins dotting the landscapes of thought and of nature, these landscapes themselves are the ruins, thoroughly ruined, pulverized, mixed with that which is not subject to decomposition on the time scales appropriate to finite existence, merged with one another and with the rest of the wanderers who roamed them. We can no longer wander among the ruins also because the interstices, the spaces between them, have been obstructed in the suffocating massiveness of the fall (the storm, the flood . . .) and because we have no ground to tread upon, the earth having fallen along with the rest. In *The Decline and Fall of the Romantic Ideal*, F. L. Lucas describes Romanticism as an "intoxicated dreaming."[1] After the fall, the relation is inverted: it is toxicity that dreams shards of words and things, the fragments of us, plants and animals, the elements, thoughts, and desires into being, which is, at bottom, unbeing.

We cannot revert to philosophical Romanticism, either. The ruins of past systems of thought, from Ancient Greece and Rome to Chinese traditions, from Gnosticism to the Indian Vedas, are not the magnificent shipwrecks of bygone worlds; they are the debris that will have crystallized in ontological toxicity. The truth of Platonism (note that I am not saying "of Plato") is depleted uranium. The historical realization of

Confucianism is the war on nature that hides the sky above Beijing from the city's inhabitants or drops it on them with the soup of chemicals that air has turned into. This is not to say that philosophical exertions from around the world be better erased from school and university curricula or from cultural awareness. On the contrary, they deserve revisiting in a patient search for the intellectual lineages of the present dump and for the countercurrents challenging its dominant drive.

Cross-cultural conversations in philosophy, notably around the question of world-destruction, are yet to mature. Such conversations are rare, for they require if not proficiency in, then intimacy with the two or more traditions that would participate in them: the ability to slide in and out of languages, frameworks for thinking, histories (including the histories of reception), contexts. What is becoming common, alternatively, is a massive tossing together of ostensibly related themes garnered from all over the world in a philosophy that is *of* and *for* the dump, even and especially under the cover of inclusivity. Approaches to "the conceptualization of spirituality at the end of life"[2] in Indian philosophy, for example, are fraught with problems embedded already in how the theme is formulated. To adhere to the logic of the concept without paying heed to other, not necessarily formal or formalizing, ways of thinking operative in the Indian context is willy-nilly to accept a Eurocentric blueprint for philosophical reflection. Ironically, the concept can twist and turn until it begins to resemble its other—the counterconcept—in a philosophical dump (*sarma*).

Besides such flagrant machinations, what are the actual possibilities of dump philosophy? How to think when the space and time for thinking and being have been drastically diminished to the point of disappearing altogether? What might a parallel to the other-than-world be within the ambit of thought?

In the first instance, we may learn a great deal from the narrator of Edgar Allan Poe's short story "A Descent into the Maelström." Written in 1841, this text served in the twentieth century as a model for understanding political processes and the Information Age by the likes of Norbert Elias and Marshall McLuhan.[3] In Poe's story, a ship is swallowed by a gigantic vortex in the ocean off the Norwegian coast. His shock and fear giving way to curiosity, the narrator decides to observe, with admirable diligence, his new and dangerous surroundings, the maelstrom he and his companions are drawn into. He recounts:

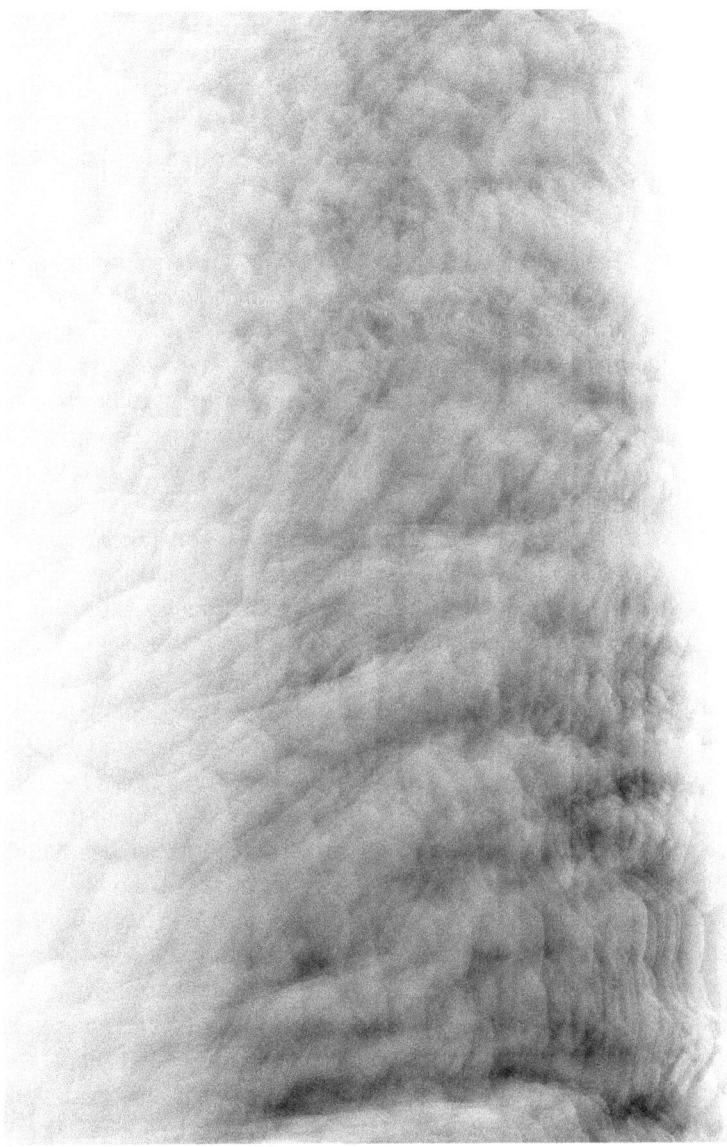

Figure 18 Drowning in, Anaïs Tondeur, 2018–20, Pigment print on Murakumo paper, 42 × 63 cm

"Looking about me upon the wide waste of liquid ebony on which we were thus borne, I perceived that our boat was not the only object in the embrace of the whirl. Both above and below us were visible fragments of vessels, large masses of building-timber and trunks of trees, with many smaller articles, such as pieces of house furniture, broken boxes, barrels and staves."[4]

As in a dump, the narrator is falling together with a mass of objects that are at first wholly undifferentiated, a part of "the wide waste of liquid ebony" pulling everything down. Without much hope for salvation, he attends to the maelstrom that is now his extreme environment and perceptually individuates object fragments in the watery abyss. Since the mutilated entities visible in an aquatic environment must be capable of floating, they are either made of wood or are the unprocessed parts of vegetation itself. So, Poe's narrator is spiraling down in abrupt, jerky, though still circular motions, together with the key ingredients of biomass, all the while conscious of his descent. His activity corresponds to our *dumpology*: coming up with the strategies for discerning the indiscernible, sorting through the dumped, rediscovering differences in the course of the massive outpouring of objects *qua* mass.

Singling out portions of the rubble in the maelstrom, which makes one's head spin, is not an easy task. Its successful accomplishment would mark a major victory in the fight against distraction and the ice-cold indifference of the dump. Thereafter, an observer may piece together the behavioral patterns of individuated fragments and, more broadly, construe an environment in the thick of environmental destitution. This is what the character in Poe's story undertakes once his horror and despair subside: he studies his unstable, swirling, watery milieu and makes deductions. "I made, also, three important observations. The first was, that as a general rule, the larger the bodies were, the more rapid their descent;—the second, that, between two masses of equal extent, the one spherical, and the other of any other shape, the superiority in speed of descent was with the sphere;—the third, that, between two masses of equal size, the one cylindrical, and the other of any other shape, the cylinder was absorbed the more slowly."[5]

Physically absorbed into the maelstrom, the narrator broaches it by observing its goings-on; he wedges some space between himself and the whirlpool. That space is a property of thinking, which, while tracking the gigantic swirl, moves against it. The renaissance of thinking in the maelstrom is a vector that would be welcome in the global dump, as well. And yet, the descent Poe depicts has a couple of advantages over the dump: (1) it happens in a vortex with a clear center and (2) it frames the narrator among, above, and beneath the still perceptible shapes of damaged objects. In his observational capacity, he can pass from a comparison of masses ("between two masses of equal size") to that of material forms ("the one spherical, and the other of any other shape"). Precisely such a passage is jammed in the decentred global dump, where all beings—organic or inorganic, human or vegetal, manufactured or grown—are stripped down to their heaviness, to matter reduced to mass.

Given the limitation we have just lighted on, and *in the second instance*, dumped thought may morph into the thinking of biomass. After the shapes of objects in the world have been mangled and vaporized in keeping with reductive knowledge protocols and business practices, after the world itself has been disarticulated, there is no compelling reason to adhere to the principles of figuration in thinking. Phenomenology has become sadly unworkable, downright inoperative in the face of this onslaught. But the historical defeat of phenomenology, as the last great legatee of Aristotelian hylomorphism, does not entail the inevitable ascendency of abstraction, of mental activity unfigured and disfigured. Since tattered ideas are dumped alongside carbon emissions, monocultures, plastic packaging, and parts of mass-produced plants or animals, thinking cannot stay far behind the overall thrust of biomassification. It must dedicate or rededicate itself to life and to mass, to life's protracted evanescence in mass, in weight, in a formless mess, in heaviness, density, and gravity.

Lest this suggestion seem fanciful, we ought to recall, together with Jean-Luc Nancy, that Latin "etymology relates thinking [*pensée*] to weighing [*pesée*]. *Pensare* means to weigh, to appraise, to evaluate (and also to com-pensate, to counterbalance, to replace or exchange)."[6] What's more, Nancy is intensely interested not only in the weighing thought but also in the weight of a thought, the title of his succinct text. For him, that weight is the weighing activity of thinking, which explores gravitational fields of meaning. Kantian philosophy stands, consequently, in need of revising: "Bodies are heavy. The weight of a localized body is the true purely sensible apriori condition of the activity of reason: a transcendental aesthetics of gravity [*pesanteur*]."[7]

Sympathetic as I am to it, Nancy's proposal falls short of the task, according to which it is vital to rethink thinking itself in the age of dumped biomass. Thought fragments tumble in a hodgepodge of body parts, disturbed elements, and the derivatives of energy production. What Nancy calls "the activity of reason"—a thinking actively weighing, comparing, evaluating, and judging the masses that are foreign to it—is deactivated, so long as the remnants of thought itself weigh us down and drop on par with other ingredients of the global dump. If thinking has no foothold outside whatever it weighs, the life (*bios*) of the mind falling with the massive wreckage of the measured and the appraised, then the a priori conditions of possibility that lay out the infrastructure for transcendental aesthetics, analytics, and, hence, for Nancy's upgraded version of the Kantian critique of pure reason as a whole, stop making sense. Further, where anything can "com-pensate" for and replace anything else in the equal devaluation and commensurability of the dumped, the need for thinking that essentially weighs heterogeneous values and meanings is obviated. Indifferent and undifferentiated, the dump neutralizes the power of thinking-weighing.

To recover the capacity for thinking, it is indeed necessary to resurface from an all-pervasive blur and discern differences again. But the retrieval of difference need not latch onto particular physical shapes, following the example of the narrator in Poe's "A Descent into the Maelström." Delving beneath the level of visibility instead, appreciating distinctions between masses, thinking must also recoil back to itself as a weighing, a dispensation to each of what is her, his, or its own. A long way from a dispassionate, "objective" assessment, the thinking-weighing of biomass will have to feel the heft of lives reduced to mass, and to feel itself feeling this heavy load, that is, to sense its weight, too, as an appendage to the one it bears. Paul Celan's famous poetic line, "The world is gone, I have to carry you [*Die Welt ist fort, ich muss dich tragen*]"[8] merits a literal interpretation: after the end of the world and upon the dissolution of life in mass, thinking can do no more than subtend the weight of dumped lives that are added to its own bulk. The *you* I have to carry is the load I support so that *you* would not continue falling alone, as a spent fragment of a fragment, tumbling even in the afterlife (and biomass is the afterlife of life itself). On my back, in my arms, or in my thoughts, I must transport you—the mass of you, no longer singular but also not amorphously general—interrupting the dumping throw in the emergent solidarity of the fall.

Where biology courses into physics, one is urged to choose between a mechanical model of causality and chaos lacking cause-effect relations. The metaphysics of causes that induce movements other than mechanical—the movements corresponding to life's generation, growth, metamorphosis, and decay—has been long discredited and ridiculed. Still, in the transition from biology to physics, between life and mass, there is a crevice that, neither physical nor metaphysical, is propitious to thought. Nothing but vacuum from the perspective of scientific rationality, this crevice is the loadbearing point for the entire system, the point at which weight can come to itself, become itself, reigniting the stalled movement of becoming.

The thinking-weighing of biomass has no theme and lacks a proper object. And yet, it thinks. "Of biomass," it is at the same time about and deriving from that of which it is.[9]

Dump Philosophy has aspired, after a fashion, to contribute to the thinking of biomass. Its argument does not content itself with taking the basic principles of physics at face value: dumping drops huge quantities of refuse down and up, to the earth in keeping with the law of gravity and to the sky with the smoke that rises from the incinerated biomass as much as from fossil fuels. The massive throw tends erratically everywhere, playing havoc with our sense of direction and thus, also, with thinking (recall Kant's question about *orientation in thought*). If, for all that, it is the fate of biomass to burn and if, moreover, what burning does to the bodies it breaks down is what the mind does to the problems it analyses, as Novalis has it, then biomass is the primary locus of elemental thinking in our "age." At stake in this thinking at the stake are the prospects of being ablaze that would fertilize the future with its ashes, rather than bring all subsequent becoming to an end. Mind and fire, problems and matter, are no longer merely analogous; they are one and the same. In the complex of biomass, how to navigate the throw and the fall while decreasing the oppressive weight of thought on other massed lives *I* have to carry? How to achieve the sort of rarefication that has little to do with dematerialization, that cherishes breathing spaces in the cluttered mess, and that lets them grow, expand from within? How to make falter and, perhaps, suspend global dumping right after it has gained its extraordinary momentum?

a poetic appendix: elemental laments

Michael Marder

water

murmuring,

flowing,

slowly growing into a wave,

rushing,

flushing sewage into the ocean,

crushing onto a shore,

carrying ships

along with a toxic mix of polyethylene, mercury, other indissolubles
and unpronounceables,

dripping with acid,

stagnating,

festering,

igniting wars over rivers and lakes,

pestering with mosquito-borne disease,

gripping the earth in floods and

triggering landslides,

making up whatever lives

and dies

earth

underlying,

drying,

flying in the air (dust),

supporting,

dwarfing the deporting authorities and us,

nourishing plants and

receiving the perishing,

cracked by tremors,

fracked by greed,

furrowed by agriculture,

burrowed by the mole and the worm,

cork tree and birch roots and weeds,

granting an afterlife to compost,

thwarting the pretense of stability,

storing natural gas, fossils, old lithium batteries,

depleted uranium, bronze coins, and everything else it conceals in
   its bowels

for the time being

air

enveloping,

developing weather fronts,

hosting birds, insects, and jumbo jets,

proving indispensable for breathing,

easing the force of gravity in proportion to being rarified,

teasing with invisibility,

lending itself to sight under a veil of smog,

suffocating,

deflating our ideologies of unlimited industrial progress.

tending to heat up in the age of global roasting,

boasting quality indices and standards,

spirited away,

expired

fire

raging,

engaging all,

ravaging forests and towns,

dispensing light and warmth,

commencing the end of the world,

resuming life,

consuming life,

cooking existence on platters of sacrifice and

ice-cold nuclear flames,

reducing to ash,

inducing frenzy,

commemorating,

incinerating the last trace

the elements

polluted,

looted,

diluted with chemicals,

analyzed into molecules,

realized as renewables,

smuggling back the infinite,

struggling to absorb our infernal waste,

sublime,

sublimated,

nearly eliminated (elementated?),

magnificent,

lamentable,

magnificently lamentable,

significantly untenable

in the haste

of the status quo to

illustrations

note on the images

Our first collaborative endeavor is a still ongoing herbarium of fragmented consciousness composed of texts and rayograms of plants growing in the radioactive soils of Chernobyl. There, we reflect on Chernobyl's Exclusion Zone as a symbol of humanity's self-exclusion from the world and of reducing the world's habitable part to almost nothing.

In our new book, the photographs themselves are created as and from the overflow of pollutants, shaping our contemporary reality, bodies, and minds. Within the manifold interactions of atmospheric flows and anthropogenic emissions, the series of *Carbon Black_Cruz Quebrada* directs our gaze to the sky above the working room of Michael Marder as he writes what will become the manuscript of *Dump Philosophy*. Simultaneously, Anaïs Tondeur turns her camera toward this strip of the changing sky. Seen through the window, which is kept as a frame, clouds stretch in atmospheric fluxes above sea currents. In the lower third of the image, human activities unfold: a man standing at the back of a van passes by; a monumental cruise ship obstructs the horizon line as it leaves the bay; a helicopter traverses the sky in a flashlight. Later the sky is apparently bereft of any human presence, while containing the carbon particles that these activities, and others like them, left behind.

The photographed skies are the ones that provided the philosopher and the artist with air. The anthropogenic particles that have for the past decades and even centuries flooded the sky travel over hundreds of kilometers gliding on atmospheric currents, from the most industrialized territories to the wildest islands and the white crust of the Arctic where they play a significant role in ice melting. This particulate matter knows no geographical borders, nor limits between inside and outside. They penetrate the membranes of our lungs, reach the deepest folds of our brains, flow with our blood cells, and trigger an increasing number of deaths. Remnants of the incomplete combustion of coal, lignite, heavy oil, or biomass, carbon black particles circulate in and around us, perhaps even carrying the virus named SARS-CoV-2.

Two years prior to Covid-19 sanitary measures, Michael Marder wore masks during the thirteen days of the artistic protocol to collect the carbon black particles that filled the sky (or the *airdump*) above him. Anaïs Tondeur later extracted the particles from the fibers of the masks with the collaboration of the physicist Jean-Philippe Putaud (JRC, European Commission) so as to transform them into ink, in a process inspired by Indian ink-making, which also uses soot, a collateral form of carbon black. The series of photographs *Carbon Black_Cruz Quebrada* are, thus, printed in part with ink made from the particles that have been collected from the masks. The spectral dust of our industrialized societies becomes constitutive of the photographs themselves and invites us, through their materiality, to witness the invisible, which shapes our intimate lives as much as our devastated world.

The second series of photographs in this book was created in a process of overlaying and multiplying the images of The Monolith, a monument erected in The Vigeland Park (Oslo). The monument, witnessed by Michael Marder during an academic visit to the city in early 2018, is composed from the intertwinement of one hundred and twenty human bodies. In the accumulation and superimposition of the same detail of the sculpture, the image itself is turned into a dump. The lines of mineral bodies melt into indiscernible matter, which blurs the boundaries between organicity and the inorganic, life and death.

As for the placement of the photographs throughout the book, at times, the images are randomly scattered and, at other times, they are dropped into a pile. In this way, the spatial arrangements of the photographs enact the paradigm of the dump with its "mechanics" of a massive and aleatory falling and with the subsequent heaping of elements that remain unarticulated among themselves.

Anaïs Tondeur and Michael Marder
Spring 2020

notes

preface: dumped

1 To be fair, it is uncertain that the elements have been ever able to appear as what they are.

2 Whether or not, in the future near or far, this book is translated and published in some of the languages I speak or think in, the titles I envision for it are: in French, *Je suis biomasse, ou la philosophie dans le dépotoir*; in Portuguese, *Para o lixo com a filosofia!*; in Spanish, *El vertedero filosófico*; in German, *Müllhaldenphilosophie*; in Russian, К философии свалки.

globality

1 The plural form of the question is not accidental. In a dump, I do not live; we live, or carry on something resembling acts of living, a "we" without togetherness, neither sharing in difference nor aired among ourselves, in the interstices between us. "I," in turn, am biomass, a massified and massed life.

2 We arrest our comprehension in its tracks as soon as we reduce movement to mechanical *kinesis*. With movement as locomotion or mere displacement conquering every sense of kinetic unfolding, our comprehension stops in its tracks; it, too, ceases moving, growing, changing forms so as to be or become more adequate to what it tries to comprehend, or even decaying and nourishing new growths with its own decomposition. These senses of movement are, incidentally, the ones Aristotle puts on par with locomotion. So, the cipher of living as movement, in movement, is incomplete without these variations, themselves organized along "active" and "passive" dimensions.

3 Santiago Zabala explores the theme of postmetaphysical being as the leftovers (or "the remains") of being in his *The Remains of Being: Hermeneutic Ontology after Metaphysics* (New York: Columbia University Press, 2009).

4 Friedrich Nietzsche, "Thus Spoke Zarathustra," in *The Portable Nietzsche*, ed. Walter Kaufmann (London and New York: Penguin, 1982), p. 417, translation modified. For more on my reading of this line and its reception by Heidegger, see my *Heidegger: Phenomenology, Ecology, Politics* (Minneapolis: University of Minnesota Press, 2018), esp. chapter 5, "Devastation."

5 For more on de-vastation, see chapter 5 of my *Heidegger.*

6 Think back, for instance, to the philosophy of Emmanuel Levinas, with its two main titles *Totality and Infinity: An Essay on Exteriority* (1961) and *Otherwise than Being, or Beyond Essence* (1974).

7 Thomas Sherratt and David Wilkinson, *Big Questions in Ecology and Evolution* (Oxford: Oxford University Press, 2009), p. 133.

8 I write "combined with" (and, accordingly, interrelated) because, taken separately, indifference to a singularity *qua* singular and thus incommensurable can be a wonderful thing, mindful of the other's transcendence, whereas care for the undifferentiated may help its undeveloped potentialities along.

9 That is the core insight of Husserl's notion of intentionality, the idea that consciousness is always a consciousness of . . ., tending toward that of which it is conscious.

10 Martin Heidegger, *Ponderings II-VI, Black Notebooks 1931–1938*, trans. Richard Rojcewicz (Bloomington and Indianapolis: Indiana University Press, 2016), p. 124.

all the world's a dump

1 In Leibniz, one finds a figuration of matter as a garden within a garden. See chapter 7 in my *The Philosopher's Plant: An Intellectual Herbarium* (New York: Columbia University Press, 2014) for more on this.

2 Cf. the epigraph. Phenomenology mandates saying what one sees. The thrust of Derrida's deconstruction is anticipated *avant la lettre*, as one says, in a written testimony to the apocalypse.

3 Claudia Baracchi, *Of Myth, Life, and War in Plato's Republic* (Bloomington and Indianapolis: Indiana University Press, 2002), pp. 107ff.

4 It bears noting that birthing is not exempt from the productivist ideology nationalism, capitalism, and instrumental reasoning jointly champion.

5 "Talk-show" is my colloquial way of translating *phenomenology*, though the order is inverted here: phenomeno-logy is, in fact, a show-talk.

6 Walter Benjamin, *Illuminations: Essays and Reflections* (New York: Schocken Books, 1968), pp. 257–8.

7 Cultural studies should be rebranded *dump studies*, a title that, far from pejorative, would give them a wide ontological scope, based on my arguments in the present study.

8 For a thought-provoking book on the subject, consult Susanna Lindberg, *Le monde défait: L'être au monde aujourd'hui* (Paris: Hermann, 2016).

mechanics: the fall, massiveness, piling up

1 Hensleigh Wedgwood, *A Dictionary of English Etymology*, 2nd edn (London: Trübner & Co., 1872), p. 230.

2 Consider a Father of the Church, Clement of Alexandria, who bridges these traditions: "the gnostic soul must be consecrated to the light, stripped of the integuments of matter, devoid of the frivolousness of the body and of all the passions, which are acquired through vain and lying opinions, and divested of the lusts of the flesh" (*Strom.* 5.11).

3 Walter W. Skeat, *A Concise Etymological Dictionary of the English Language* (Oxford: Clarendon Press, 1882), p. 182.

4 Timothy Mousseau et al., "Highly Reduced Mass Loss Rates and Increased Litter Layer in Radioactively Contaminated Areas," *Oecologia*, 175, no. 1(May 2014): 429–37.

5 I am alive to the aporia of trying to circumscribe the amorphous and will address it in what follows, especially in the chapter titled *dumpology*.

falling before and after the death of god

1 The Russian word *skotina*, meaning cattle, a beast, or a brute is derived from the Greek *skotein*.

2 The following empty space has been introduced by the word processing program on which, or with which, I have been writing this text. I wished to preserve this contingent, though also implacably necessary, feature in the "final" version.

3 Vladimir Bibikhin, *Les [The Woods]* (St. Petersburg: Nauka, 2011), p. 86.

4 St. Augustine, *Essential Sermons*, ed. Boniface Ramsey (Hyde Park, NY: New City Press, 2007), p. 120.

5 Augustine, *Essential Sermons*, p. 62.

6 Doesn't all mediation do just that—cushion the fall?

7 Joel Achenbach et al., "Why People Are Marching for Science: 'There Is No Planet B'," *The Washington Post*, April 22, 2017, https://www.washingt onpost.com/national/health-science/big-turnout-expected-for-march-for-science-in-dc/2017/04/21/67cf7f90-237f-11e7-bb9d-8cd6118e1409_sto ry.html, last accessed on February 4, 2018.

8 This "mutual adjustment" bespeaks a basic environmental justice.

je suis biomasse

1 Lauren Provost, "'Je suis Charlie': qui est à l'origine de l'image et du slogan que le monde entier reprend par solidarité," *Huffington Post (France)*, January 7, 2015, http://www.huffingtonpost.fr/2015/01/07/je-su is-charlie-origine-createur-joachim-roncin-slogan-logo-solidarite-charlie-h ebdo_n_6431084.html, last accessed on February 4, 2018.

2 Jacques Derrida, *The Animal That Therefore I Am*, trans. David Wills (New York: Fordham University Press, 2008), p. 64.

3 Michael Marder, "The Idea of Following in the Age of Twitter," *Al-Jazeera*, May 20, 2012, http://www.aljazeera.com/indepth/opinion/2012/05/2 012519123159732261.html, last accessed on February 4, 2018.

4 Could we say, in a speculative tone, that actualized revolutions, those that carry their potential through to historical actuality, relate to the world (the social, political, and economic world, but also the broader horizons of existence) as to a dump? The Terror that burst on the historical scene on the heels of the French Revolution and the Stalinist purges in Soviet Russia would then be a prime example of the insatiable aspiration to pure ideality that creates a virtual desert and a dump, where the corpses of victims are massively and indiscriminately piled up. Should this line of thinking bear fruit, the problem of "successful" revolution will require a radical reframing: How to prevent revolutionary or postrevolutionary regimes from ending up in a dump?

5 Walt Whitman, *Leaves of Grass* (Iowa City: University of Iowa Press, 2009), p. 43.

6 "There appears to be no antithesis it [Roman Catholicism] does not embrace. It has long and proudly claimed to have united within itself all forms of state and government. . . . But this *complexio oppositorum* also holds sway over everything theological." Carl Schmitt, *Roman Catholicism and Political Form*, trans. G. L. Ulmen (Westport, CT: Greenwood Press, 1996), p. 7.

7   Elena Pierazzo, *Digital Scholarly Editing: Theories, Models and Methods* (London and New York: Routledge, 2015), p. 67, FN#3

8   Friedrich Nietzsche, *On the Genealogy of Morality*, trans. M. Clark and A. J. Swensen (Indianapolis and Cambridge: Hackett Publishing, 1998), p. 58.

9   Nietzsche, *On the Genealogy of Morality*, p. 58.

10 Any wonder, then, that the state-*form*, and most of all the social protections guaranteed by the welfare state, are on the wane?

11 Novalis, *Philosophical Writings*, ed. and trans. Margaret Mahoney Stoljar (Albany, NY: SUNY Press, 1997), p. 48.

12 G. W. F. Hegel, *The Science of Logic*, trans. George Di Giovanni (Cambridge and New York: Cambridge University Press, 2010), p. 67.

13 G. W. F. Hegel, *Encyclopedia of the Philosophical Sciences in Basic Outline, Part I: Science of Logic*, trans. and ed. Klaus Brinkmann and Daniel O. Dahlstrom (Cambridge and New York: Cambridge University Press, 2010), p. 143.

14 Gerardo Ceballos et al., "Biological Annihilation Via the Ongoing Sixth Mass Extinction Signaled by Vertebrate Population Losses and Declines," *PNAS* 114, no. 30 (July 2017): E6089–E6096.

antilogos

1   Aryeh Finkelberg, *Heraclitus and Thales' Conceptual Scheme: A Historical Study* (Leiden: Brill, 2017), p. 60, FN #55.

2   This distinction should not be confused with passive and active voices, for example, the Spinozan *natura naturata* and *natura naturans*. For more on this notion of energy, refer to my *Energy Dreams: Of Actuality* (New York: Columbia University Press, 2017).

3   For more on this dissociation, see my *Pyropolitics: When the World Is Ablaze* (London: Rowman & Littlefield Int'l, 2015), esp. ch. 2.

4   Nietzsche, "Thus Spoke Zarathustra," p. 129.

5   Martin Heidegger, *Introduction to Metaphysics*, trans. Gregory Fried and Richard Polt (New Haven, CT and London: Yale University Press, 2000), p. 142, translation modified.

6   Jacques Derrida, *Specters of Marx: The State of the Debt, the Work of Mourning, and the New International,* trans. Peggy Kamuf (London and New York: Routledge, 1994), p. 161.

toward an intellectual history of heaps, piles, and other jumbled things

1 William James, *The Principles of Psychology*, vol. I (Mineola, NY: Dover, 1950), p. 488.

2 James, *The Principles of Psychology*, p. 488.

3 James, *The Principles of Psychology*, p. 488.

4 Edmund Husserl, *Formal and Transcendental Logic*, trans. Dorion Cairns (The Hague: Martinus Nijhoff, 1969), p. 60.

5 Husserl, *Formal and Transcendental Logic*, p. 61.

6 All references to Kant's *Critique of Pure Reason* appear in brackets, containing the pagination of the first (A) or second (B) editions. The English translation used is Immanuel Kant, *Critique of Pure Reason (The Cambridge Edition of the Works of Immanuel Kant)*, ed. and trans. Paul Guyer and Allen W. Wood (Cambridge: Cambridge University Press, 1999).

7 All references to Hegel's *Phenomenology of Spirit* appear in brackets, containing the abbreviation *PhS* and a relevant paragraph number. The English translation consulted is G. W. F. Hegel, *Phenomenology of Spirit*, trans. A. V. Miller (Oxford and New York: Oxford University Press, 1977).

8 In Aristotelian terms, does this simply mean that the categories *this* (first *ousia*), *when* (time), and *where* (place) are shorn of meaning unless articulated among themselves and with other categories in a joint determination of their object?

our polluted senses

1 An earlier version of this chapter appeared in the *New York Times* column "The Stone" under the same title on October 9, 2017. https://www.nytimes.com/2017/10/09/opinion/light-noise-pollution-senses.html, last accessed on April 4, 2018.

2 Keith Ridler, "Idaho Hopes to Bring Stargazers to First US Dark Sky Reserve," *US News*, September 15, 2017 https://www.usnews.com/news/best-states/idaho/articles/2017-09-15/stargazers-eye-the-nations-first-dark-sky-reserve-in-idaho, last accessed on February 4, 2018.

3 On the toxicity of sugar refer to Robert Lustig, *Fat Chance: Beating the Odds against Sugar, Processed Food, Obesity, and Disease* (London and New York: Penguin, 2013), pp. 257ff.

4 Maurice Merleau-Ponty, *Phenomenology of Perception*, trans. Colin Smith (London and New York: Routledge, 2004), p. 11.

5 Aleksandr Pushkin, "Pir vo vremya chumy," in *Collected Works in Ten Volumes*, ed. D. Blagoy et al., vol. IV (Moscow: State Literary Publisher, 1960), p. 378, translation mine.

toxicity

1 Julian Cribb, *Surviving the 21st Century: Humanity's Ten Great Challenges and How We Can Overcome Them* (Cham: Springer, 2017), p. 201.

2 I owe this insight to Doreen Mende.

3 Mario Burger and Henri Slotte, "United Nations Environment Programme Results Based on the Three DU Assessments in the Balkans and the Joint IAEA/ UNEP Mission to Kuwait," in *Depleted Uranium: Properties, Uses, and Health Consequences*, ed. Alexandra C. Miller (Boca Raton, London and New York: CRC Press, 2007), p. 250.

4 Terry Eagleton, *On Evil* (New Haven and London: Yale University Press, 2010), p. 16.

5 —— "Rising Global Cancer Epidemic," *American Cancer Society*, https://www.cancer.org/research/infographics-gallery/rising-global-cancer-epidem ic.html, last accessed on February 4, 2018.

6 Rachel Carson, *Silent Spring*, 40th Anniversary Edition (Boston and New York: Houghton Mifflin, 2002), pp. 34-5.

7 For more on the clandestine theology of water, see Kimberly Patton's, *The Sea Can Wash Away All Evils: Modern Marine Pollution and the Ancient Cathartic Ocean* (New York: Columbia University Press, 2006).

8 John Sallis, *Logic of Imagination: The Expanse of the Elemental* (Bloomington and Indianapolis: Indiana University Press, 2012), p. 147.

shitty apocalypse, or scatological eschatology

1 —— "Sanitation Factsheet," *World Health Organization*, http://www.who.int/mediacentre/factsheets/fs392/en/, last accessed on February 14, 2018.

2 Kerstin Magnusson et al., *Microlitter in Sewage Treatment Systems: A Nordic Perspective on Waste Water Treatment Plants as Pathways for*

Microscopic Anthropogenic Particles to Marine Systems (Copenhagen: Nordisk Ministerråd, 2016), p. 29.

3 Barry M. Peake et al. *The Life-Cycle of Pharmaceuticals in the Environment* (Sawston and Cambridge, UK: Woodhead Publishing, 2015), p. 234.

4 Works of art, Freud claims, are the sublimations of shit. The toilet, in turn, presupposes a toiling activity. Hence, also, Marx's point in the third volume of *Capital*: "Under the heading of production we have the waste products of industry and agriculture, under that of consumption we have both the excrement produced by man's natural metabolism and the form in which useful articles survive after use has been made of them." Karl Marx, *Capital: A Critique of Political Economy*, vol. III, trans. David Fernbach (London and New York: Penguin, 1991), p. 195.

5 Consult Jacques Derrida, "Biodegradables: Seven Diary Fragments," trans. Peggy Kamuf, *Critical Inquiry*, 15 (Summer 1989): 812–73.

6 The same is true for foods wrapped in plastic. While isolating a food item from dirt, the wrapping pollutes it by imparting parts of its own chemical makeup to whatever it wraps.

7 Here and thereafter, "SE" refers to the Standard Edition of Freud's writings: *The Standard Edition of the Complete Psychological Works of Sigmund Freud*, 24 Volumes, trans. and ed. James Starchey (London: Vintage, 2001).

8 It is high time for a dialectical critique—coming on the heels of a traditional Marxist response to existentialism—of the abstract, self-indulgent, ultimately bourgeois subject of self-questioning. What is unforgivable is carrying such critique out unconsciously and with recourse to the objective residua of industrial activity that absorb the figure of *anthropos* and its lived temporality into the Anthropocene.

9 Keep in mind that we are dealing with the second childishness here.

falling in love and being dumped

1 And how does this stand (or fall) when it comes to falling in love with the earth? Or to dumping in?

2 Julia Kristeva, *Powers of Horror: An Essay on Abjection*, trans. Leon Roudiez (New York: Columbia University Press, 1982), p. 6.

3 Vladimir Solovyev, *Smysl lyubvi* [*The Meaning of Love*] (St. Petersburg: Azbuka-Klassika 2016), pp. 105–6, translation mine.

4 Maurice Merleau-Ponty, *L'Oeil et l'Esprit* (Paris: FolioPlus, 2006), p. 9.

on the arcane utility of the useless

1   Robert Willig, "Economic Effects of Antidumping Policy," in *Brookings Trade Forum: 1998*, ed. Robert Z. Lawrence (Washington, DC: Brookings Institution Press, 1998), p. 59.

2   Georges Bataille, *The Accursed Share: An Essay on General Economy*; Volume I: Consumption, trans. Robert Hurley (New York: Zone Books, 1991), p. 25.

3   Bataille, *The Accursed Share* I, p. 25.

4   Bataille, *The Accursed Share* I, p. 26.

5   Brink Lindsey and Daniel Ikenson, *Antidumping Exposed: The Devilish Details of Unfair Trade Law* (Washington, DC: Cato Institute, 2003), p. 2.

6   Karl Marx, *Capital: A Critique of Political Economy*, vol. I, trans. Ben Fowkes (London and New York: Penguin, 1976), p. 276.

7   Lindsey and Ikenson, *Antidumping Exposed*, p. 1.

8   Marx, *Capital* I, p. 792.

9   Marx, *Capital* I, p. 788.

10 Marx, *Capital* I, p. 813.

11 "Hunger will tame the fiercest animals, it will teach decency and civility, obedience and subjection, to the most perverse. In general, it is only hunger which can spur and goad them [the poor] on to labor; yet our laws have said they shall never hunger. . . . Legal constraint is attended with much trouble, violence, and noise . . . whereas hunger is not only peaceable, silent, unremitting pressure, but as the most powerful natural motive to industry and labor, it calls for the most powerful exertions." Quoted in Karl Polanyi, *The Great Transformation: The Political and Economic Origins of Our Time* (Boston: Beacon Press, 1957), p. 113.

12 On the distinction between housing and dwelling, see my *Heidegger: Phenomenology, Ecology, Politics*.

the portrait of a thing as its own wastebasket

1   An earlier version of this chapter appeared in *Cabinet Magazine* (60) under the titled "Being Double" in the Winter 2015–16 issue.

2 Jean-Luc Nancy tacitly contradicts Heraclitus when he contends that "everything reverts to the general equivalence in which one sleeper is worth as much as any other sleeper and every sleep is worth all the others, however it may appear . . . Sleep itself knows only equality, the measure common to all, which allows no differences or disparities" [*The Fall of Sleep*, trans. Charlotte Mandell (New York: Fordham University Press, 2009), p. 17]. From whose perspective does this equivalence reveal itself? How does sleep know equality, "the measure common to all"? And is this commonality or, rather, the formal and indifferent equality of the dump?

3 Tamsin Walker, "Bottled Water Not Safe from Microplastic Contamination," *Deutsche Welle*, March 14, 2018, http://www.dw.com/en/bottled-water-not-safe-from-microplastic-contamination/a-42936246, last accessed on March 15, 2018.

4 Luce Irigaray, *To Be Two*, trans. Monique Rhodes and Marco Cocito-Monoc (London and New York: Routledge, 2001), p. 12.

5 "[O]nce he [the primitive] arrived at the idea that man is a body that a spirit animates, then he must of necessity impute to natural bodies that same sort of duality, plus souls like his own. . . . [T]he phenomena of the physical world above all—the movement of the waters or of the stars, the germination of plants, the abundant reproduction of the animals, and the rest—are accounted for by the soul of things." Emile Durkheim, *The Elementary Forms of Religious Life*, trans. Karen Fields (New York: The Free Press, 1995), p. 50.

6 Such senselessness does not impede economic logistics exploiting this defect and profiting from it.

dumpology

1 Immanuel Kant, *Critique of the Power of Judgment (The Cambridge Edition of the Works of Immanuel Kant)*, Revised edn, trans. Paul Guyer and Eric Matthews (Cambridge: Cambridge University Press, 2001), p. 271.

2 Here, I am not referring to that which is *worth* saving; the themes of value and of the value of value are, in this respect, derivative.

estamira, esta mira, "this sight"

1 —— "Best 8 Landfills in Washington, DC with Reviews," https://www.yellowpages.com/washington-dc/landfills, last accessed on March 13, 2018.

2   *Classified* is one of those curious speculative words in the Hegelian tradition that unsays or contradicts itself: classified information is set apart from the rest, withheld, secretive; the classified section of a newspaper presents a range of categories of goods and services to the public gaze, making known, promoting, calling attention.

3   https://www.wm.com/store/dumpster-rental/small-business-landing.jsp ?cmp=IYP_YP_N, last accessed on March 13, 2018.

4   Thanks to Marcia Sa Cavalcante Schuback for calling my attention to this film.

5   This and all subsequent translations of Estamira's monologues are mine.

6   For more on the vitality of edges, consult Edward S. Casey's outstanding philosophical works, particularly the recent *The World on Edge* (Bloomington and Indianapolis: Indiana University Press, 2017).

the writing dump

1   Literature anticipated the preeminence of the code with its *litera*, alphabetic letters.

2   I myself have been a proponent of granting an existential-phenomenological "world status" to plant life.

3   One of these is Markus Gabriel's 2017 study *Why the World Does Not Exist* (London and New York: Polity).

4   I am much obliged to Wendy Lochner for raising this thought-provoking possibility.

parts of the void

1   Fernando Pessoa, *The Book of Disquiet*, trans. Margaret Jull Costa (London: Serpent's Tail, 1991), p. 118.

2   Hans-Christoph Askani, *Das Problem der Übersetzung—dargestellt an Franz Rosenzweig* (Tübingen: J. C. B. Mohr, 1997), p. 210.

3   Talmudic commentary then reads cosmogony through the lens of Jewish history: "R. Simon b. Lakish interpreted the verse *And the earth was formless and void* as applying to [the various] exiles [experienced by the Jews]." Wilfred Shuchat, *The Creation According to the Midrash Rabbah* (Jerusalem and New York: Devora Publishing, 2002), p. 96.

4   Gershom Scholem suggests that Bar Hiyya was the first to identify *tohu* with *hulē* and *bohu* with *morphē*, or form. For more on this, refer to Scholem's *Origins of the Kabbalah* (Princeton, NJ: Princeton University Press, 1987), p. 62.

5 Hannah Arendt, *The Human Condition*, 2nd edn (Chicago and London: The University of Chicago Press, 1998), p. 144.

6 Arendt, *The Human Condition*, p. 143.

7 Arendt, *The Human Condition*, p. 9.

8 Alex Hern, "Bitcoin Mining Consumes More Electricity a Year than Ireland," *The Guardian*, November 27, 2017, https://www.theguardian.com/te chnology/2017/nov/27/bitcoin-mining-consumes-electricity-ireland, last accessed on March 13, 2018.

9 Simone Weil, *Waiting for God*, trans. Emma Craufurd (New York: Harper & Row, 1973), p. 111

in-formation

1 With the expression "world interior" I am implicitly citing Rilke's *Weltinnenraum*, which Peter Sloterdijk picks up nearly a century after the coinage in his 2013 book *In the World Interior of Capital: For a Philosophical Theory of Globalization*, trans. Wieland Hoban (Cambridge, UK and Malden, MA: Polity, 2013), p. 197.

2 D. M. Dominoni et al. "Clocks for the City: Circadian Differences between Forest and City Songbirds," *Proceedings of the Royal Society B,* 280, 1763(2013): 20130593; Roy Chepesiuk, "Missing the Dark: Health Effects of Light Pollution," *Environ Health Perspect*, 117, no, 1 (2009):A20–7.

3 Shoshana Zuboff, *The Age of Surveillance Capitalism: The Fight for a Human Future at a New Frontier of Power* (New York: Public Affairs, 2019), p. 68.

4 I resort to this little-used term so as to designate a level below Michel Foucault's micropolitics.

rameau's nephew for the twenty-first century

1 Denis Diderot, *Rameau's Nephew and First Satire*, trans. Margaret Mauldon (Oxford, UK and New York: Oxford University Press, 2006), pp. 3–4.

2 Diderot, *Rameau's Nephew*, p. 29.

3 Diderot, *Rameau's Nephew*, p. 7.

4 Diderot, *Rameau's Nephew*, p. 15.

5 Diderot, *Rameau's Nephew*, p. 33.

6 Diderot, *Rameau's Nephew*, p. 46.

7 Hannah Arendt, *The Origins of Totalitarianism* (Cleveland and New York: Meridian Books, 1964), p. 409.

8 Diderot, *Rameau's Nephew*, p. 73.

9 Diderot, *Rameau's Nephew*, p. 76.

10 Diderot, *Rameau's Nephew*, p. 26.

11 Diderot, *Rameau's Nephew*, p. 6.

12 Diderot, *Rameau's Nephew*, p. 12.

13 Diderot, *Rameau's Nephew*, p. 7.

14 Diderot, *Rameau's Nephew*, p. 8.

15 Diderot, *Rameau's Nephew*, p. 8.

16 Diderot, *Rameau's Nephew*, pp. 58–9.

17 Diderot, *Rameau's Nephew*, p. 84.

18 Diderot, *Rameau's Nephew*, p. 20.

dump philosophy, or the task of thinking in the age of dumping

1 F. L. Lucas, *The Decline and Fall of the Romantic Ideal* (Cambridge: Cambridge University Press, 1936), p. 46.

2 Hamilton Inbadas, "Indian Philosophical Foundations of Spirituality at the End of Life," *Mortality* (2017), DOI: 10.1080/13576275.2017.1351936.

3 I am grateful to Aurelién Gamboni for discussing this text with me.

4 Edgar Allan Poe, *Complete Tales and Poems* (Edison, NJ: Castle Books, 2002), p. 65.

5 Poe, *Complete Tales and Poems*, p. 66.

6 Jean-Luc Nancy, *The Gravity of Thought*, trans. François Raffoul and Gregory Recco (Atlantic Highlands, NJ: Humanities Press, 1997), p. 75.

7 Nancy, *The Gravity of Thought*, p. 77.

8 Paul Celan, *Breathturn into Timestead: The Collected Later Poetry*, Bilingual edn, trans. Pierre Joris (New York: Farrar, Straus & Giroux, 2014), p. 97.

9 In this sense, too, it resonates with plant-thinking.

Index[1]

---

[1]Despite the alphabetical order it follows, this index is a dump of sorts, or a reflection of the dump. Do the seemingly random juxtapositions of words in it not begin the work of "undumping"? Consider: antidumping legislation and Antigone; *bios* and by-products; compost and *conatus essendi*; deindividuation and democracy; eschatology and Eve; Gnosticism and growth; heap and heaven; information and innocence; jointure and justice; Klee and *kosmos*; literature and litter; microcosm and microplastic; *National Geographic* and *natura*; obsolescence and ontology; plastic and Plato; quantity and *quodlibet ens*; refuse and relation; Silicon Valley and sin; transcendence and trash; uranium and utility; *viriditas* and void; wealth and weight; (Mao) Zedong and *zōe*…